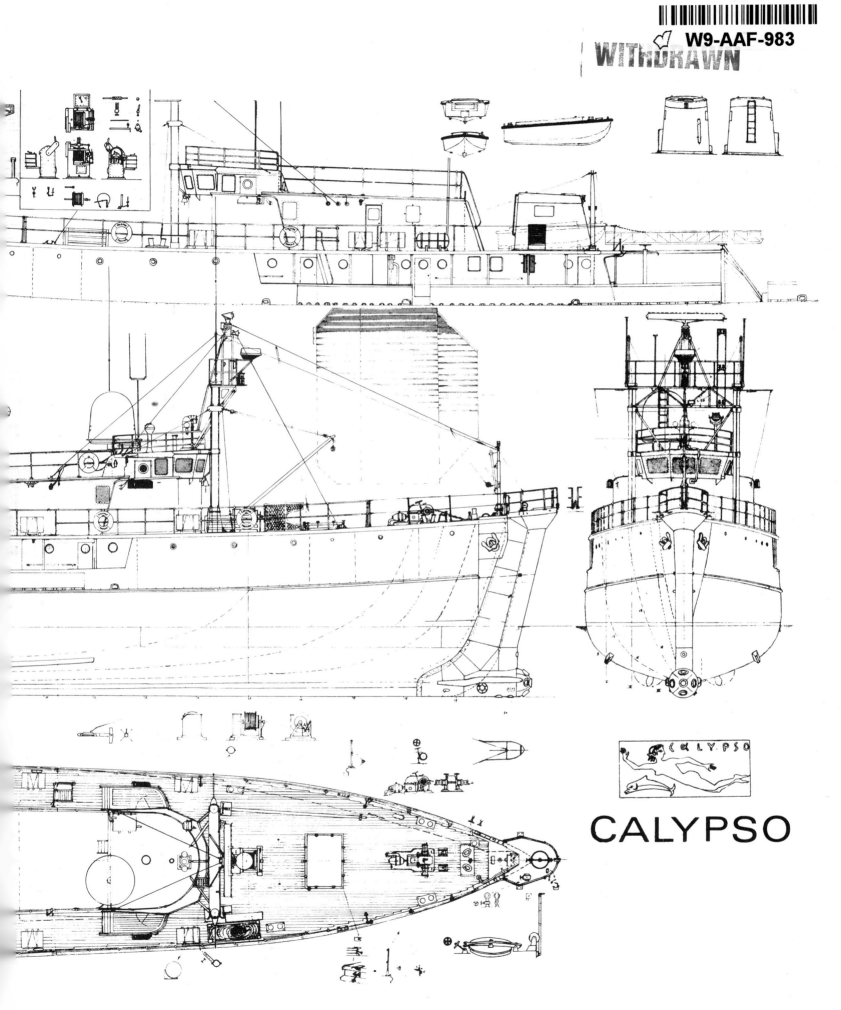

CALYPSO

JACQUES COUSTEAU'S CALYPSO

JACQUES COUSTEAU'S
CALYPSO

By Jacques Cousteau and Alexis Sivirine

HARRY N. ABRAMS, INC., PUBLISHERS, NEW YORK

Deep in the Mediterranean lies this delicate bouquet of marine
animals made up of zoanthids, sponges, and corals. The zoanthids
stretch out their tentacles—which release tiny harpoons that
inject a paralytic toxin—to capture a potential meal.

First page: a diver waits at decompression level, using the
hull of *Calypso* to prevent himself from surfacing too rapidly.

Preceding pages: while the helicopter hovers above, *Calypso*
prepares to launch the diving saucer in front of a cliff of ice.

On the back jacket: a red and white starfish, photographed in
the Indian Ocean.

Editor: Robert Morton
Editorial Assistant: Beverly Fazio
Picture Researchers: Anne-Marie Cousteau, Nancy Grant
Designer: Darilyn Lowe

Library of Congress Cataloging in Publication Data

Cousteau, Jacques-Yves.
Jacques Cousteau's Calypso.

Includes index.
1. Calypso (Ship)   2. Oceanographic research.
3. Underwater exploration.   I. Sivirine, Alexis.
II. Title.   III. Title: Calypso.
VM453.C67   1983   551.4'6'00723   83-3751
ISBN 0-8109-0788-7

© 1983 Jacques-Yves Cousteau

Published in 1983 by Harry N. Abrams, Incorporated, New York

Printed and bound in Japan

The hot air balloon rises majestically over the Antarctic expanse, providing a high vantage point and a stable camera platform.

Contents

This medusa, or jellyfish, travels by jet propulsion through the waters off the coast of southern California.

1933 . . . I am a young midshipman aboard the training vessel *Jeanne d'Arc* on a hydrological exercise in the Bay of Port Dayot, Vietnam. A local Vietnamese fisherman guides our launch. As I watch him dive beneath the boat, I am filled with wonder. At noon, the sea dead calm, the heat stifling, he slips naked into the water without gear or goggles of any kind and disappears without a ripple. He surfaces a minute later with a marvelous fish wriggling in each hand and explains with a mischievous smile: "They nap at this time of day."

1936 . . . Between two sea duties, I am teaching aboard the battleship *Condorcet* in Toulon. One of my students tells me of an uncle of his who goes diving off the French Riviera with Tahitian goggles and fishes for groupers, dorados, and leerfish with a bow and arrow.

These two incidents fire my imagination. There beneath the keels of our boats lies a little-known yet penetrable universe teeming with life—a wild marine jungle separated from our civilized world only by the surface of the sea, an ever changing boundary that conceals the world below from our eyes and has enveloped her in mystery and legend.

From that time on, my friends Philippe Tailliez, Frédéric Dumas, and I talk of nothing but exploring the sea. We visit the pioneers in the field, Commander de Corlieu and Commander Le Prieur. We try out existing equipment—and make some of our own—goggles, fins, mouthpieces, spears (which we soon abandon), cameras, and oxygen tanks (with which I later have two serious accidents).

War comes and finally the Armistice, during which time Emile Gagnan and I develop the aqualung and I make my first two films, *Par dix-huit mètres de fond* and *Epaves*, awarded a prize at the Cannes Film Festival.

Thanks to these two films, I succeed in convincing the Naval Chief of Staff, Admiral Lemonnier, to commission a new Undersea Research Group in Toulon. Later I become captain of our first underwater-research ship, the M.V. *Ingénieur Elie Monnier*.

Our programs take us from research in diving physiology to minesweeping; from exploring the Fountain of Vaucluse to the first tests of the bathyscaphe. But as varied as these tasks are, I always dream of exploring the seas throughout the world . . . and in spite of the sage advice of my elders I decide to attempt the impossible and plunge into the Great Adventure. It is this adventure on board *Calypso* that is meticulously retraced in the pages that follow.

J.-Y. Cousteau

Jacques-Yves Cousteau studies a flight of shore birds on an Indian Ocean island in 1967.

A hogfish and a blue tang seek refuge in an underwater cave off Mexico's Yucatan coast.

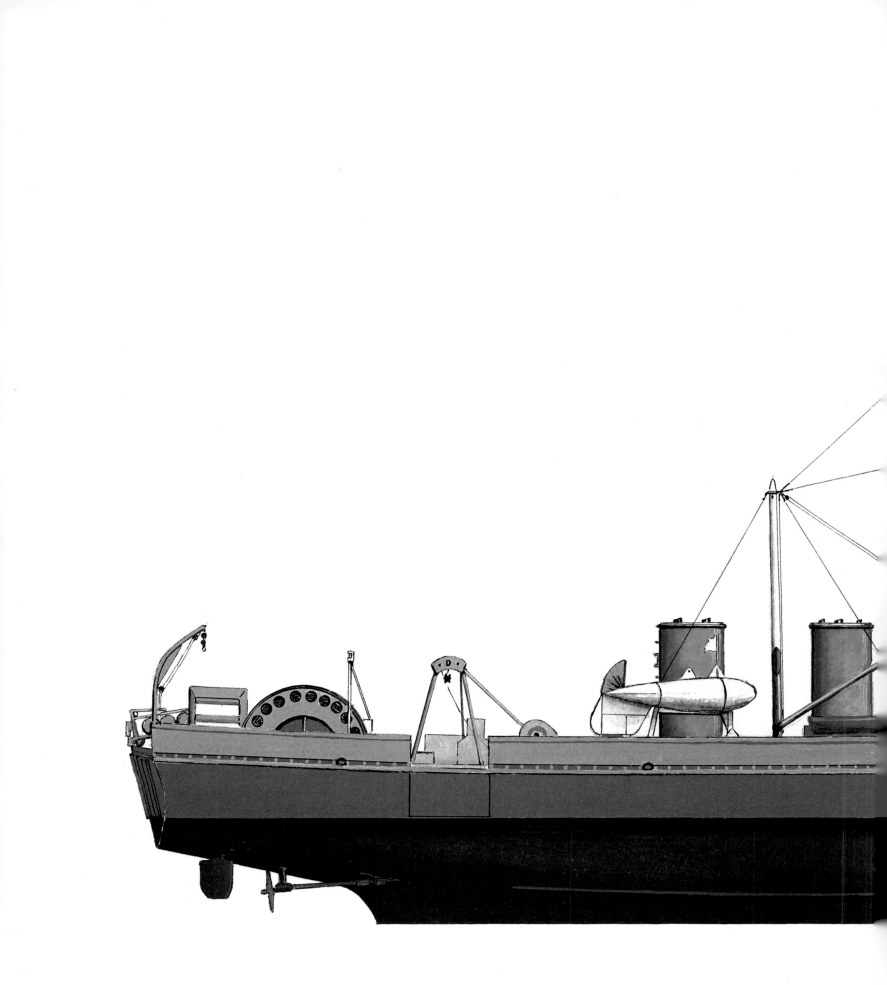

1942—Minesweeper J-826, built in the United States, goes to war in the British navy.

1950—Newly named, *Calypso* ("water nymph") does service in Malta as a coastal ferry.

1954—Under Cousteau's guidance, *Calypso* proves herself a first-class research vessel.

1958—A new silhouette and new gear modify the sturdy craft.

1959—Special radio equipment is added for a topographic mapping project.

1968 — Calypso returns to her native country for the first time.

On *Calypso*'s bridge Captain Cousteau peers into the hood of
a radar screen to locate a sunken ship in the Red Sea.

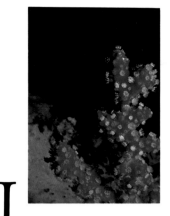

J acques-Yves Cousteau had dreamed of devoting his life to exploring the mysteries of the ocean world long before acquisition of the ship *Calypso* made such exploration feasible. Throughout his twenty-year career with the French navy he worked toward making his dream a reality by diving underwater on his own, experimenting with existing diving equipment, and, finding much of this equipment insufficient to his lofty ambitions, creating improved tools to meet his needs.

Most notable of Cousteau's early inventions was the aqualung, or SCUBA (Self-Contained Underwater Breathing Apparatus), which he designed in 1943 with Emile Gagnan. The aqualung made possible for the first time extended exploration of the depths of the sea; without it, divers' fields of activity were severely limited. Cousteau's appointment in the 1940s to captain the *Ingénieur Elie Monnier*, the world's first marine-research vessel, in the service of the French navy's newly founded Undersea Research Group, provided him with the opportunity to study firsthand all the equipment necessary for a ship completely dedicated to undersea research.

Soon Lieutenant Commander Cousteau decided the time was right for him to pursue his dream actively, and in 1950 he obtained leave from the navy to devote himself full time to exploring the ocean. His first task, of course, was to find a ship. The ideal craft would be solidly built, easy to handle, and have a shallow enough draft to let her maneuver in and out of tight spots with a minimum of effort. His search for such a vessel led him to André Auniac, then head of the naval shipyards in Antibes. Auniac, in turn, put Cousteau in touch with Noel Guinness, an Englishman who shared the captain's passion for the sea. Guinness, recognizing that Cousteau was possibly the only man alive who could make large-scale underwater exploration a viable reality, offered to buy and outfit a vessel.

CALYPSO IS FOUND

Cousteau found his ideal ship in a naval surplus yard in Malta. Formerly a World War II minesweeper, built for the British navy in the United States in 1942, *Calypso*—whose name means "water nymph"—had since been employed as a ferryboat between the islands of Gozo and Malta, but she had all the qualities for which

Encounter with Adventure

Simone and Jacques Cousteau, at right, stand with members of their improvised crew on the deck of *Calypso* during the ship's first sea trial after its purchase.

Top: The gorgonian is a plant-like structure well known to marine enthusiasts for its beauty as souvenir and decoration. Until the nineteenth century great healing powers were ascribed to all corals, and they were worn as jewelry and woven into armor to ward off evil spirits.

Cousteau had been searching. Although not very large (she measures about 140 feet in length and 24 feet across the beam), *Calypso* proved to be a solid vessel with twin engines, a double hull of wood in excellent condition, and remarkable maneuverability. The sales contract was signed in Nice on July 19, 1950, and *Calypso* was taken immediately to the shipyard in Antibes, where her conversion from ferryboat to oceanographic research vessel was begun.

Naturally, extensive alterations had to be made in order to turn this former minesweeper and ferryboat into a floating laboratory. The interior accommodations were remodeled, navigational aids were added, and special facilities for diving equipment were installed. An underwater observation chamber, known as *Calypso*'s "false nose," was also installed. This false nose was connected to a metal well built around the prow itself and extending eight feet below the waterline. The bulbous chamber had eight portholes that enabled a few of *Calypso*'s crew to see and film underwater without even leaving the ship. Another significant addition was the observation tower constructed on *Calypso*'s foredeck: the tower serves the threefold purpose of radar antenna mount, upper bridge for navigation, and crow's nest from which the larger marine animals could be observed.

Calypso thus became a floating base for marine research and exploration such as the world had never before seen. Her forte in those early days, as well as during her entire career to date, was in making the new opportunities provided by her special equipment

Sailing as a ferry out of Malta, and under a British flag, *Calypso* carried as many as 11 cars and up to 400 passengers.

available to the most diverse scientific disciplines. Besides oceanographic research as such, *Calypso* was equipped to observe biological population patterns and behavior of marine and coastal animals; to study the morphology of coral reefs; to investigate the effects of various instruments employed by divers; and to provide the scientific community with direct observations of undersea geological structures. In addition to her special diving-oriented technology, *Calypso* carried standard instruments such as corers, grabs, sampling bottles, current meters, and seismic-reflection profilers, which facilitated diverse assignments in such fields as topography, acoustics, geophysics, physics, chemistry, and geology.

All these alterations and special equipment proved to be very expensive—too expensive, in fact, to be covered even by the generous amount of money given by Noel Guinness. Cousteau, therefore, had to turn elsewhere for funding; luckily, he received donations of most of the equipment and materials from private sources, including manufacturers, and from the French navy. Thus did Cousteau's dream, shared by an entire community of young scholars, industrialists, and sailors, become a reality. *Calypso*'s mystique would grow ever greater.

Once Cousteau had solved the problems of obtaining and outfitting a suitable ship, he came to the unsettling realization that there was more to operating a floating laboratory than he had anticipated. Running the ship itself and recruiting and maintaining a crew were all extremely expensive, and required, therefore, an administrative structure. Moreover, Cousteau himself wished to remain free of these responsibilities in order to have time to pursue his other projects and writings. And above all, he wanted to establish beyond question the integrity of his venture. Therefore, following the advice of his friend Claude-Francis Boeuf, Cousteau established the COF—Campagnes Océanographiques Françaises, or French Oceanographic Expeditions—a nonprofit organization responsible for commissioning, managing, and administering funds for *Calypso*'s expeditions. Monies available to the COF were to derive solely from fees from *Calypso*'s privately sponsored research, from grants, and from royalties from films, books, articles, and television specials.

By June 1951 *Calypso* was more or less ready to put to sea. Jacques Cousteau, his wife, Simone, and Frédéric Dumas, Cousteau's friend and fellow diver since their navy days, were impatient to learn how the ship would behave at sea. They decided to run preliminary tests off the coast of Corsica. The crew, the most extraordinary that *Calypso* has ever known, was improvised by inviting a few friends. It included Roger Gary, a manufacturer; Gary's brother-in-law, the Marquis Armand de Turenne; Edmond Maruic, an architect; Pierre Malville, a restaurateur from Antibes; Jacques Ertaud, a young cameraman; a lone sailor nicknamed Málaga, from Gary's yacht; and an engineer named Octave Léandri. (The only especially trained person aboard the ship, besides Cousteau himself, of

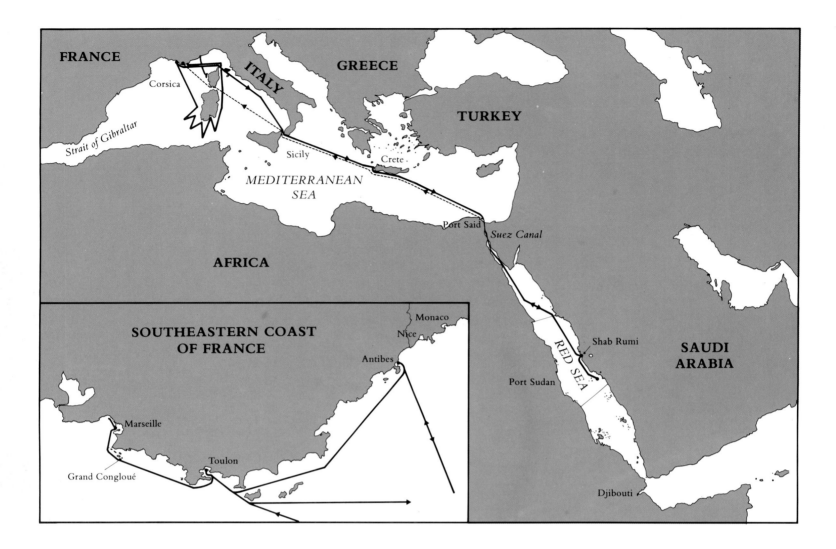

course, was Léandri, called Titi, the first man hired. He would stay with *Calypso* for more than fifteen years.) In addition, the Cousteaus' two sons, Jean-Michel, aged twelve, and Philippe, ten, were taken on as cabin boys.

Thus, almost exactly one year after acquiring *Calypso*, Cousteau set sail in her for the first time, and under the French flag—although not legally so, for the papers necessary to register her as a French vessel were not yet ready. *Calypso* herself was far from being completely equipped: there were not even any lifeboats yet, for example, so Malville had had to lend Cousteau his dinghy for the maiden voyage. *Calypso* served her apprenticeship for two weeks in preparation for her new destiny, then returned to port, where her future was laid out for her.

Her first real voyage of exploration was planned, amidst a feverish atmosphere, at her home base of Toulon on the south coast of France. An ambitious, multifaceted program was worked out. Since the COF still did not have enough funds to entirely support the ship's expeditions, Simone and Jacques Cousteau devoted a major part of their personal resources to the undertaking. The number of paid team members was reduced to the minimum, and everyone, including the scientists, was required to help with the shipboard chores and the watch at sea.

The story of day-to-day life aboard *Calypso* is not always a glamorous one. Rather, it is often the story, as it was on this first trip, of exhausting work, both underwater and aboard ship; of dives at all hours of the day and night, and in all kinds of weather; of routine daily maintenance and emergency repairs. But it is also the story of the exhilaration experienced by diving to new depths and discovering new facets of the infinite varieties of undersea life and the fascinating, often breathtaking, visual splendors of the sea. The great esprit de corps aboard ship causes everyone to pitch in to help with chores that are far beyond the scope of their professional duties. For aboard *Calypso* one thing is held in common by crew members, divers, and scientists alike: an almost single-minded devotion to the sea and to all things contained therein.

Calypso's first official skipper—François Saout, chief boatswain and an expert in long-distance sailboat racing—came to her by way of the navy, as did René Montupet, her chief engineer, and Jean Beltran, a sailor and diver. A former navy cook, Fernand Hanen, was also hired. Jacques Ertaud remained on board as a cameraman, and Jean de Wouters d'Oplinter, an engineer who would later invent the world's first amphibious camera, "Calypso-phot," was signed on as underwater photographer. Octave Léandri also stayed on to assist Montupet with *Calypso*'s two engines. Captain Cousteau and Frédéric Dumas were to supervise the diving, and Simone Cousteau assumed the stewardship. In fact, during *Calypso*'s more than thirty years of activity, Simone, of all the crew, has spent the most time aboard the ship, sailing the seas of the world and contributing more than anyone to the team's spirit of adventure and enthusiasm.

Cousteau then hired Jean-Loup Nivelleau de la Brunnière Véron to fill the position of ship's doctor. Véron had wanted so desperately to be a part of the *Calypso* team that he had presented to Captain Cousteau a letter of application which contained falsified credentials. He was to admit a month later that the letter was a tissue of lies, but by then he had proved himself so competent that words on paper no longer made a difference to his new career.

Several varying scientific disciplines were also represented in *Calypso*'s crew. A team of marine biologists consisted of Pierre Drach, professor of marine biology at the Sorbonne; Gustave Charbonnier, an assistant at the malacology (mollusk) laboratory of the National Museum of Natural History in Paris; and Claude Lévy, an assistant at the Roscoff Marine Station in Brittany. Geology and the morphology of the coral reefs were the province of André Guilcher, a professor of geology at the University of Nancy; Haroun Tazieff, a famous vulcanologist; and Vladimir Nesteroff, a geology assistant at the Sorbonne. Claude-Francis Boeuf and his assistants, Bernard Calamme, deputy director of La Rochelle Laboratory, and Jacqueline Zang, a chemist at the same laboratory, were responsible for the hydrological studies undertaken aboard *Calypso*.

The Cousteau family—Jean-Michel, Simone, Jacques-Yves, and Philippe—explore the clear depths of the Mediterranean.

Finally, on November 24, 1951, *Calypso* sailed from the Toulon arsenal and headed for the Suez Canal and the Red Sea. The crew's almost childlike enthusiasm was quickly put to the test when on the third day of the crossing, November 27, *Calypso* was caught in a violent northeast storm raging in the south Adriatic Sea. One after another the two engines failed, and the crew, becoming desperate, attempted in alarm to rig up a makeshift sea anchor to keep the ship from drifting into danger. But the "emergency" soon proved to be a false alarm—simply a case of clogged fuel filters. The engines, after two years of inactivity, had not been cleaned out completely. Soon the voyage continued.

This first exploratory mission was dedicated to observing the magnificent coral reefs of the labyrinthian Farasan Archipelago off the Saudi Arabian coast, studying the abundant marine life found there, and photographing and filming the brilliantly colored fish. All these studies were to be made firsthand, for, with more than six thousand dives to his credit today, Captain Cousteau had formulated even then what was a unique concept of oceanography. Instead of relying on shipboard instrumentation to test and analyze specimens of the undersea world, *Calypso*'s crew practiced Cousteau's personal ideology. "*Il faut aller voir*," he says—"we must go and see for ourselves." This almost militant insistence on the necessity of man's presence in the water to arrive at a true understanding of that world was put into application for the first time on this voyage.

The Red Sea, the destination of this mission, offered exceptional possibilities for such a program. The Farasan Bank, south of Jidda along the coasts of the Hejaz and Yemen, held a particular attraction for *Calypso*'s team. Except for Australia's Great Barrier Reef, these islands comprised the world's greatest coral complex; yet only the outer limits of the reef had already been charted. Maps of this area described its center with this challenging note: "unexplored region, studded with reefs separated by deep but non-navigable channels." This previously uncharted center was where the diving team concentrated their attention and activities; they pitched camp on the island of Abu Latt, where they established a laboratory to be shared with scientific teams of geologists, biologists, and hydrologists. Meanwhile, *Calypso* was crisscrossing the Red Sea, making soundings that proved for the first time the existence of undersea basins of volcanic origin which contained extraordinary strata of valuable minerals. These basins would later be studied under the name "hot brines."

The expedition was a marvelous revelation to the entire group of divers, scientists, and crew members. Cousteau's philosophy of direct observation, first put into effect here, proved invaluable to the scientists, who, with the aid of the divers, made many startling discoveries about the marine life of the area. Rich collections were formed of underwater photographs and sample organisms. Among these were heretofore unknown species of flora and fauna, several of

Above and below: before perfecting small, lightweight camera housings for underwater work, the Cousteau photographers fitted out steel oil drums with watertight gaskets and pressure-proof glass windows.

which were named after persons or things associated with the expedition—*Calypseus, Saouti, Cousteaui.*

Calypso returned to her home port of Toulon on February 5, 1952. The expedition had proved to be an unqualified success, but the feeling of exuberance among the crew did not last long. Immediately upon their arrival at port they learned of the untimely deaths of fellow team members Claude-Francis Boeuf and Jacqueline Zang, both killed in an accident at the Addis Ababa airport while returning to Europe.

Now that *Calypso* had proved the value of Cousteau's approach ("*Il faut aller voir*"), it was decided to apply this philosophy to a new activity: underwater archaeology. Having already had the opportunity to explore many sunken ships off the coasts of Provence and Tunisia, both Captain Cousteau and Frédéric Dumas felt that systematic excavation of sunken ships had much to offer in the way of naval archaeology. They chose for their first subject a ship-

ARCHAEOLOGY UNDER THE WAVES

For entertainment, the diving team established on Grand Congloué island for the winter put on a mock-Greek banquet in imitation of the society whose relics they were rescuing from the sea floor.

wreck found about twelve miles from Marseille, along the southern coast of Grand Congloué, an inhospitable island, arid and ridiculously small—little more, in fact, than a large barren rock.

In July 1952 *Calypso* headed for Vieux Port, the harbor of Marseille, which the crew adopted as their temporary home port. *Calypso* shuttled back and forth between Marseille and Grand Congloué for the next six months, setting things up as a base for the most important underwater excavation to date. Because *Calypso* could not remain moored by the island all winter long without risk, a hut equipped with electrical generators, furniture, and various accessories was set up on the rocks of Grand Congloué. Thus "Port

Calypso" was established. Five or six divers, who were regularly supplied with fresh provisions, lived on Port Calypso that winter. Among these divers was Raymond Coll, then a young man of sixteen, who would grow to be one of the best divers on *Calypso*'s team.

During these months the men of *Calypso* made dozens of dives each day to explore the wreck, which, from evidence they gathered, was determined to be a Roman ship dating from the third century B.C. To free the wreckage from the sediment and debris that had accumulated over the centuries, the divers employed on this mission a new device: an air lift—a cumbersome and often capricious piece of equipment made specially for underwater archaeology.

The air lift consisted mainly of a huge suction hose based at an air compressor on the island, supported on a long boom stretching out from shore and dropping 150 feet straight down to the bottom of the sea near the wreck. Divers would guide the mouth of the hose into the sea floor sediments to suck up anything there. When working properly, the air lift would spew tons of material from the sea bottom—sand, shells, pots, tools—into a large sieve-like bin ashore. Sand, mud, marine organisms, and small shells would pass through the mesh of the strainer, leaving behind only the larger objects, which usually included pieces from the shipwreck. The hose needed constant attention, however, for it would often clog up, causing it to writhe uncontrollably in the water, whipping to and fro and escaping from the divers' hands. Restraining the pipe was so strenuous an activity that as a rule each pair of divers was limited to two fifteen-minute sessions each day while operating the air lift.

Thousands of amphoras, pieces of ancient Campanian pottery, and other artifacts were brought to the surface, inventoried, and classified by *Calypso*'s team. The excavation of this two-thousand-year-old shipwreck aroused considerable interest throughout the world. People from all walks of life filed aboard the ship to glimpse the many wonders restored from the depths of the sea; the commander of this French military region was taken down on a dive; and several new divers became attracted to *Calypso*, joining her team. Albert Falco, who would soon become *Calypso*'s chief diver, signed on as a permanent member on September 22. With him, Henri Goiran, Raymond Kientzy (almost always called Canoë), André Laban, and Jean-Pierre Servanti now formed the nucleus of *Calypso*'s diving team—around Frédéric Dumas. A spirit of high enthusiasm pervaded the group.

The good feelings would not last long, however, for tragedy struck the expedition. As Jacques Cousteau arrived on *Calypso* at Grand Congloué on November 6, 1952, he noticed that the mooring buoy to which the ship generally tied up had drifted about half a mile away, a result of a storm the previous night. Jean-Pierre Servanti, one of *Calypso*'s best and most experienced divers, dove below to check out the situation. Upon reaching the sea's floor, he discovered that the chain attached to the buoy was broken, and the

Supervised by Captain Cousteau, a diver enters the water off Grand Congloué to operate the television camera being lowered from *Calypso*'s rear boom.

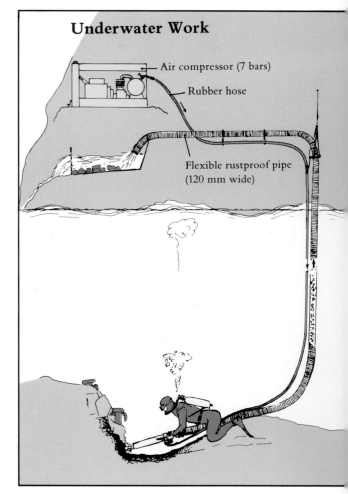

Underwater Work

Air compressor (7 bars)

Rubber hose

Flexible rustproof pipe (120 mm wide)

A diagram of the suction dredging system with its air compressor, intake tube, and recovery sieve.

A diver maneuvers the snout of the suction tube.

23

anchor lost. *Calypso*'s many divers searched for the anchor all day, but to no avail, so Servanti proposed following the furrow left in the sand by the chain as it drifted. Captain Cousteau agreed reluctantly; he worried about the depth of the dive, shown on the echo sounder as 230 feet. Servanti descended anyway, and all went well until his air bubbles, which were being traced aboard ship, suddenly disappeared. Albert Falco dove into the water instantly, followed closely by Ertaud and Yves Girault. They found Servanti lying unconscious on the bottom. The three divers brought their motionless comrade to the surface, where, after many attempts at resuscitation, he was placed in *Calypso*'s recompression chamber as the ship sped back to Marseille. Despite five hours' treatment aboard *Calypso* and in the large recompression tank at the Marine Firemen's Station in Marseille, Servanti could not be brought back to life.

The whole *Calypso* team were deeply grieved. Still, the expedition could not be forgotten, and diving was resumed. More artifacts were brought on board, then sent to the Musée Borély in Marseille. The recovery work was well advanced and the valuable material could now be turned over to other specialists for cataloguing, study, and publication. Finally, on January 23, 1953, *Calypso* returned to her home base in Toulon for a period of much-needed rest and servicing.

For both the Red Sea and the Grand Congloué expeditions the *Calypso* team had devised and performed underwater tasks so innovative that they continually found themselves needing special equipment, tailored to their specific requirements. In nearly every case such equipment did not already exist. Therefore, to fill the gap between need and availability, Cousteau and a group of Marseille officials established on March 4, 1953, a second nonprofit organization, the purpose of which was to conceptualize and develop prototypes of revolutionary marine instruments. Called the Office Français de Recherches Sous-Marines (OFRS), this foundation began modestly, but soon expanded. Eventually, it designed and built almost all the equipment aboard *Calypso*, and in 1968 changed its name to the Center for Advanced Marine Studies, its French acronym CEMA. Many other marine research organizations would benefit from the efforts of CEMA.

In April 1953 *Calypso* returned to Marseille to resume work on the archaeological dig off the island of Grand Congloué. Underwater television equipment—one of the first such systems in the world and, of course, developed by the OFRS—allowed archaeologists to follow the divers' activities while they themselves remained on board the ship. The system's advanced technology also made it possible to televise even in very turbid water.

By July, after nearly a full year of work at Grand Congloué, *Calypso*'s divers had made more than 3,500 dives to depths of over 130 feet. They retrieved well over ten thousand pieces of ancient pottery—which greatly enhanced the collection of the Borély Museum as well as the reputation of *Calypso* and her crew.

Albert Falco takes the lead in offloading some of the 2200-year-old amphoras recovered at Grand Congloué.

Dr. Harold Edgerton, crouching at left, displays an array of his underwater flash and camera gear.

The crew's first stop after Grand Congloué was Antibes, where *Calypso* took on board a new team member, Harold Edgerton, a professor at the Massachusetts Institute of Technology and an expert in deep-sea photography. Edgerton's invention of the first electronic flash earned him the nickname Papa Flash from the other members of the ship's crew. Papa Flash's equipment greatly expanded the range of undersea photography. Now divers could take pictures of many marine animals and plants in the great, dark depths native to them, rather than having to wait until sample organisms were captured and brought to the surface, as had previously been the case.

In August 1953 *Calypso* sailed again, this time on a voyage to find traces of the owner of the vessel wrecked at Grand Congloué—a voyage that would take the ship some five thousand nautical miles throughout the eastern Mediterranean. It had been discovered during the excavation that the wrecked ship had set sail from the Greek island of Delos in the year 209 B.C. Furthermore, the amphoras retrieved from the wreck had engraved on their necks the initials "S.E.S.," together with images of dolphins and anchors, determined to be the marks of a Greek shipowner of the period named Markos Sestios. In the attempt to find out more about the shipowner, the *Calypso* team visited in succession the Greek ports of Kithira, Andikithira, and Delos, where a villa was identified, through its mosaics, as having indeed belonged to Sestios. Having thus proved his identity to their satisfaction, *Calypso*'s crew returned to Marseille on September 2, then continued their excavating work at Grand Congloué.

Their archaeological duties were interrupted on September 21, however, when *Calypso* was called upon to aid in the rescue of the Italian ship *Donatello*, which had run aground and broken up on the rocks of nearby Riou Island in calm but foggy weather. Finally, in November of 1953, two years after it had begun, *Calypso*'s archaeological mission at Grand Congloué was completed.

It was never *Calypso*'s destiny to remain immobile at a work site; her destiny was to travel the seas, exploring their depths and discovering their many mysteries. The crew was anxious to move on. The ship returned to Marseille, where she was equipped with a new radar unit and a modern precision echo sounder. From there she went back to her home port of Toulon to prepare for her next mission.

Calypso sailed from Toulon on January 7, 1954, on a new mission: to prospect for oil in the Persian Gulf on behalf of the D'Arcy Exploration Company, a subsidiary of British Petroleum.

The course she steered was by now a familiar one: to Port Said, through the Suez Canal, and on to the Red Sea, where she dropped anchor just off the Daedalus Reefs. After reconnoitering the coral islands of Yemen, *Calypso* was caught in a violent storm in the Strait of Bab al Mandab and was forced to seek refuge in Djibouti. Inspection by divers revealed that the ship had been severely damaged by the storm; the starboard anti-roll keel was in very bad shape and needed to be repaired immediately. *Calypso* made her way to Aden on the southern coast of the Arabian peninsula for repairs. The floating dry dock located there was just large enough to accommodate her. While *Calypso* was in Aden, Wallace Brown, a Canadian geophysicist, and Alan Russell, an Australian geologist, came aboard and prepared to supervise her upcoming oil-prospecting program.

After the necessary repairs were completed, *Calypso* continued on her mission. Briefly stopping at Al Mukalla and Muscat, she made a colorful anchorage in Elphinstone Inlet in the Strait of Hormuz, a narrow bay enclosed by high banks, which *Calypso*'s crew characterized as the hottest place on earth. Finally, the divers thoroughly examined the area of Abu Dhabi, the site of the mining concession that had been granted to British Petroleum. In the course of this work, *Calypso* made four hundred stations or anchorages.

These anchorages were set up often under the most difficult conditions, for it was the season of the *khamsin*, a hot, raging southerly wind more violent than even the mistral, that strong, dry northerly wind common in southern France. The optimum location for each station was accurately determined with the help of a Decca radio navigation network installed on some nearby islands. The purpose of these many anchorages was to map the substrata of the

PROSPECTING FOR OIL

sea floor by taking density readings with a gravimeter at each station, and thus compiling a picture of the earth's crust that might reveal pockets where oil could likely be found.

Divers also descended in an attempt to take geological samples from the sea floor, but the bottom was composed of rock so hard that even the heavy steel drop-corers could not penetrate it—the steel pipe returned to the surface as pleated as an accordion. It was found necessary for the divers to pierce the rocks with pneumatic drills that shook them unmercifully. Compounding the difficulty was the fact that the waters were filled with sharks. The divers descended in an antishark cage, but were forced to leave its safety to do the actual drilling and to collect their samples. Yet although sharks circled around continuously, they proved not as dangerous as the poisonous underwater snakes, about seven feet long, which slithered between the bars of the cage. Nevertheless, this dangerous and exhausting work, first tested on this expedition, led to the discovery and exploitation of one of the richest oil sites in the world.

Her job completed, *Calypso* left Abu Dhabi for Bahrain and then Doha, on the Qatar Peninsula, where she put into port to fill up with water and fuel. While at Doha, the crew paid a visit to the *Shell Guest,* a ship that remained permanently at anchor in the shallow bay to serve as quarters for the technical staff of the Shell Oil Company. And good news reached *Calypso* at Doha: the Ministry of National Education had signed on April 1, 1954, an agreement with Cousteau's original organization, the Campagnes Océanographiques Françaises (COF), ensuring financial aid for *Calypso.* The Ministry of National Education would underwrite a major part of the cost of future expeditions, and for the next several years *Calypso* would act as an official French oceanographic vessel.

Calypso was now ready to set out on new explorations, and she was joined for them by two new passengers, an old friend, Gustave Charbonnier, of the National Museum of Natural History in Paris, and James Dugan, an American writer. On her way back from the Persian Gulf, *Calypso* called at several stops in the Indian Ocean: Dennis Island and Mahé in the Seychelles, Providence Island, and Diégo-Suarez (Antseranana) on the tip of Madagascar. Finally she arrived at the island of Aldabra, northeast of the Comoros, a little-known coral and mangrove sanctuary.

Several channels cut through the thick coral ring at Aldabra, leading to a large lagoon where, it was said, a German cruiser lay up during World War I. On Aldabra the famous land turtles, relatives of those found on the Galápagos Islands in the Pacific, wandered freely over the mangrove-covered land. These turtles were truly enormous; many weighed well over three hundred pounds. Camp was pitched ashore so that the scientists could study the island's ground fauna, which included a unique wingless bird. Meanwhile, the divers were busy collecting many different marine specimens in the nearby coral reefs.

Next on *Calypso*'s agenda after Aldabra and after her return to

For centuries sailors have celebrated first crossings of the equator with costume parties and prank-filled initiation ceremonies. Here, King Neptune, surrounded by his court, hears a proclamation.

the Mediterranean was the Gulf of Lions on Corsica, where, from September 15 to September 25, biological dredgings and samplings were made with the help of a team of scientists from the Arago Laboratory. Then, for the entire month of October, the crew studied the geology of the Corsican coasts with Professor Bourcart. Finally, on November 1, 1954, the ship and the crew—equally exhausted—both returned to Toulon for the winter repair period, and then, on January 27, 1955, to Marseille, where *Calypso* was put into dry dock until the beginning of March.

Work does not stop while the ship is in dock, however. Analyses and evaluations of the data collected on the last expedition and preparations for the next take up much of the crew members' time. In addition, the OFRS, of course, works all through the year.

During this particular layover, the OFRS was busy perfecting, with the help of *Calypso*'s scientific and diving crew, the first diving saucer, a device that would revolutionize underwater exploration.

While diving in the outer reefs of the Farasan Archipelago in the Red Sea, Cousteau had found that all the steep cliffs of the reefs stopped at around 150 feet at a kind of beach, which he believed corresponded to a fossil sea level during a recent glacial period. The beach then sloped gently down to about 200 feet, where a brilliant-colored second cliff dropped steeply off into the dark blue depths of the sea, inaccessible to divers. Obsessed by the desire to explore the cliffs as far down as 1,000 feet, Cousteau began to muse about a device, resistant to pressure and easily maneuverable, that would allow him to extend his field of investigation. Discussing his idea at lunch one day in *Calypso*'s mess, Cousteau picked up two soup plates and, placing the rims one against the other, indicated the shape he wanted his device to have. The OFRS immediately went to work on his idea, and that same year made a model of an ellipsoid that would suit Cousteau's purpose. The model, built on a one-tenth scale, measured eight inches in diameter and six inches high, and did indeed greatly resemble Cousteau's two soup plates.

In 1955 the scale model passed a compression test at the Toulon arsenal. The device, in this preliminary stage, was called the "Tortoise," alluding to the original submarine invented by the American David Bushnell as well as to its turtle-like shape. Not until the full-sized vehicle was completed, some four years later, would it be called the "diving saucer."

While *Calypso* was in dry dock for the winter of 1955, preparations were being made for her next voyage, during which the crew were to shoot their first feature-length film, a documentary to be called *The Silent World*. Much needed to be done during these winter months; cameras, film, medicine, buoys of all kinds, and sundry tools were gathered in preparation. All this was done under the watchful eyes of Frédéric Dumas and the filmmaker Louis Malle, who, at twenty-one, was on his way to achieving great fame as a movie director.

FILMING IN THE SILENT WORLD

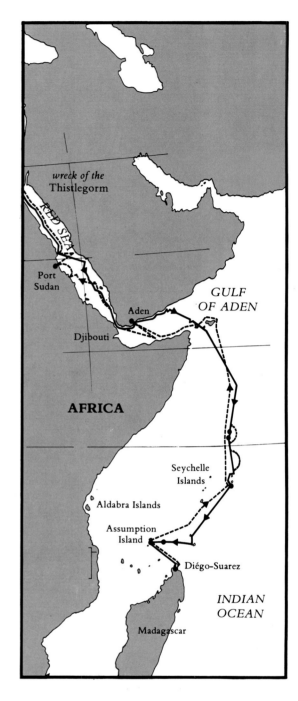

On March 8 *Calypso* set sail on a voyage that would take her over twelve thousand nautical miles. She arrived in Port Said after one week, having gone through a violent storm en route. Continuing her voyage, the ship traveled down the Red Sea, stopping at the wreck of a ship at Ras Muhammad. The wreck, which had obviously been there for some time, was determined by the echo sounder to be at a depth of 100 feet; its metal hull had broken open and huge pearl oysters were encrusted on it. The ship was identified by its bell as the *Thistlegorm*, commissioned in Glasgow and sunk during World War II by German fighting planes.

While investigating the wreck, the divers were suddenly confronted by an enormous fish. It so startled Frédéric Dumas and Albert Falco that they pulled back and sought safe haven in one of the ship's narrow passageways. "He was as big as a truck," said Dumas later; Falco added, "His scales were as big as my hand." This "truck-fish" (as it was then called) was also seen the next day by Louis Malle, and later identified as a monstrous wrasse, probably weighing over a ton. No record existed of this common fish having grown so large, but however incredible it seemed, the encounter was well verified.

Reefs and islands gave way to one another as *Calypso* continued on her way through the Red Sea and into the Indian Ocean, toward the Seychelles. While traveling in the doldrums area near the equator and in a dead-calm sea, the ship came across an immense pod of sperm whales. Several groups of the huge cetaceans surrounded *Calypso*. Then a baby sperm whale, just sixteen feet long and probably no more than a few weeks old, perhaps mistaking the bulk of the ship for its mother, brushed alongside *Calypso* and got caught in the port propeller. The propeller sliced the infant, injuring it severely and causing profuse bleeding. Divers' attempts to rescue the whale were halted by the gathering of countless sharks, which, drawn by the smell of blood, were already circling the stricken creature. Unable to help the whale, Cousteau and several cameramen descended in the antishark cage to film the terrifying spectacle they knew would ensue.

The first sharks to gather, *Carcharhinus longimanus*, or "long-finned sharks," were soon joined by a lone blue shark, about fifteen

André Laban and Captain Cousteau prepare to descend in the antishark cage.

feet long, with a long, pointed snout. The sharks continued to circle the bleeding whale for an hour or so, then warily began prodding it with their snouts, barely grazing it, making no attempt to bite. Then, suddenly, the blue shark lunged and attacked; instantly, the others followed suit, rushing in a frenzy at the unfortunate infant, and savagely tore it to bits. This was the first time such a scene had been captured on film, and *Calypso*'s crew had photographed it from a distance of only a few feet.

Witnessing this violent performance was a draining experience for the entire crew, so they took some time off on Mahé in the Seychelles to unwind. Soon *Calypso* was on her way again, though, for after her short break she left for Aldabra via Farquhar Island, Gambetta Island, and Assumption Island. Assumption Island in the Aldabran group proved to be a genuine paradise for the cameramen. Besides the striking natural scenery, many interesting animals were discovered there. Giant green turtles were filmed both underwater and while laying their eggs on the beach. The divers tamed a large grouper (christened Ulysses), which later became a famous "screen star."

The cameramen could have occupied themselves at Assumption Island for a much longer time, but the monsoon had begun to blow, and so *Calypso* made her way back north, through very rough seas. Finding better weather, the crew stopped the ship near the coast of Somalia, where they spent some time filming dolphins, porpoises, and several species of whales. *Calypso* then again proceeded northward, into the Red Sea, stopping often to photograph the beautiful coral reefs and the radiant fish indigenous to them.

By June 12, *Calypso* had once again reached Ras Muhammad and the site of the *Thistlegorm*, the crew's favorite sunken ship, and the most beautiful of all the wrecks the divers had ever inspected. The ship provided the cameramen with a long sequence of film—some of the most striking footage contained in *The Silent World*.

After entering the Mediterranean, *Calypso* spent a few days off the coast of Crete, diving with Greek and Turkish sponge fishermen clad in traditional helmet suits, and filming them at their work, then continued toward home. Finally, on June 27, 1955, the ship returned to Marseille, where she would receive the repairs that were necessary after her strenuous journey.

Her rest was a short one, however, and for the next four months *Calypso* periodically sailed on voyages throughout the eastern Mediterranean, stopping at Greece, Istanbul, Crete, and Bizerte on the North African coast. Much was accomplished during these brief sojourns: under the direction of Professor Henri Lacombe, the crew set up hundreds of hydrographic stations to measure the temperature and salinity of the water; a new winch outfitted with 16,400 feet of steel cable enabled them to lower Edgerton cameras to the bottom of the Matapan Trench, almost three miles down, the deepest trench in the Mediterranean; and many different species of marine life were collected and identified.

Anatomy of CALYPSO

Technical features of Calypso

The ship's features have changed considerably in the course of various renovations. In 1982, *Calypso's* features were the following:

Hulls
Overall length: 139 feet
Beam: 25 feet
Average load draft: 10 feet
Maximum height above sea level: (radar) 37½ feet
Light displacement: 324 tons
Full load displacement: 402 tons
False nose with 6 observation portholes: 10 feet deep underwater
Two rudders

Speed
10 knots with 2 motors running at 900 rpm
7 knots with 1 motor running at 800 rpm

Auxiliary motors
2 GM 6.71 generators
—6-cylinder motor—140 hp—1,200 rpm
—60kw, 110v DC generator
1 GM 2.71 generator
—2-cylinder motor—40 hp
—20kw, 110v DC generator
5 40 hp and 25 hp Johnson outboard motors for the launches and Zodiacs
3 portable Oran generators

Storage tanks
Gas-oil

4 tanks in afterhold	847 cubic feet
10 ballast tanks	812 cubic feet
Total	1,659 cubic feet

Daily consumption when using a 60kw generator: 950 gallons
Range with 2 days' reserve: 11 days at a rate of 10 knots: 2,640 miles
For longer trips, Calypso loads 52-gallon tanks on deck.

Oil tanks
2-tank engine—370 gallons

+ 132 gallons	502 gallons
4 tanks of 53 gallons per engine	212 gallons
Total	714 gallons

4 drums of various oil for the crane, the winch, outboard motors, and for the Junkers' air compressors.
Gas tanks for outboard motors (quarterdeck): 2 disposable tanks of 118 gallons each
Kerosene tanks for helicopter: 6 disposable drums of 53 gallons each

Fresh water

4 aft tanks	1,850 gallons
Ballast tanks	3,170 gallons
Total	5,020 gallons

A distiller can produce 317 extra gallons daily.
2 Allied Water Systems desalination units giving 900 gallons of fresh water per day

Air compressors
2 Junkers compressors—4 stages
Power: 30 hp; output 132 g/h; type 4FK115 air at 205 bars; 4 storage tanks 264 gallons each; Feeding ramp for 10 bottles.
Ingersoll-Rand MP air compressor
Bauer air purification system for diving
Portable Bauer compressor for diving

Propulsion
2 General Motors engines type GMC 8-268A straight 8 cylinders
Two-stroke cycle diesel engine
Inlet air by lights—exhaust by valves
Pressure-charge: 300 g
Fuel-oil injection pumps
Dry geared engines with separated oil boxes and cooling system
Manual dry gearbox
Positive drive clutch and reverse
Reduction of speed 1/3 in forward position
Capacity 580 hp at 1,250 rpm
Maximum crank-shaft speed 1,250 rpm
Franklin Electric bow thruster 7.5 hp

Yumbo hydraulic crane
Power, 4 tons; equipped with a hydraulic winch; hydraulic generator operated by an electric 40 hp/11v motor; Marrel compressor: 90 bars; in case of emergency can be driven by the winches' generator

Hydraulic winch, including:
2 drums, 1 carrying 9,850 feet of 15/32-inch steel cable; the other, 3,290 feet of the same cable
This winch is also equipped with headstocks and a special wheel for nylon cable.

Duclos hydraulic generator driven by a 40 hp/110v electric motor and a variable flow pump which in turn drives a hydraulic motor and the winches

Warluzel sounding winch: 5/32-inch cable

Fore windlass with 3 hp electric motor pulling 2 anchors, 551 lbs. and 661 lbs. and 5 links of chain of 10, or 492 feet each

Aft windlass with 2.5 hp electric motor pulling a 330 lb. anchor and 5 links of chain of 10, or 492 feet

Electrical equipment: both 60kw, 110v DC generators and the 20kw generator feed many lines, of which:
2 4kwa 220v single-phased transformer
1 25kwa 220/380v 3-phased transformer
1 special radar transformer
1 Allis Chalmers 1,250w radio transformer
1 voltage regulator for television
1 Special Collins radio transformer
1 1kwa transformer for COMSAT system

Navigation Equipment
Automatic steering wheel with Hardlandic Brown automatic pilot; can be switched instantly to handwheel
Brown gyroscopic compass
Ocean Sonic OSR 119T/Raytheon DE 721 A sound transmitter
Radar Decca serial D7 type 808 with 12-in. PPI scope—transmitter maximum power: 75kw
Collins BLU receiver radio
Collins 51 J4 receiver—500 khz to 30 mhz
Emergency LMT transmitter/receiver
Fregate transmitter/receiver
TCS transmitter/receiver—1 600 khz to 4 mhz
100W Collins transmitter—500 khz to 30 mhz—150w telegraph

Sound units
Ampex stereo loud speaker
Marantz amplifiers, 2 × 50w

Television system
Video circuit with 2 waterproof Grundig cameras and 5 receivers

Photo laboratory with darkroom and air-conditioned film storage cabinet

Walk-in iceboxes and many refrigerators

Repair workshop with drills, lathes, grinding wheels, etc.

Galeazzi recompression chambers

Launch made of light alloy and 4 inflatable boats, Zodiac MKIII; 2 "Bombard" lifeboats big enough to hold 20 people each

Beside the permanent features listed above, various pieces of equipment are loaded aboard *Calypso* according to the different missions.
Air lift for sea floor excavation
Underwater gravimeter
Two-man diving saucer SP-350
Underwater scooters
SP-500 one-man minisub
Trawler troika equipped with flash to take pictures and films of the sea bottom
Hughes 300C helicopter with its pad
40- to 50-foot wide Vulcoon hot air balloons
Decca transmitting equipment
Meteorology data receiver from NASA
Equipment allowing continuous analysis of the chlorophyll content in the sea water
Apparatus to measure currents
Dredges
Coring devices to collect samples of mud and soil
Water sampling bottles
Rolling measuring device
Thumper sound transmitter to analyze sea bottom
EGG lateral sound transmitter
Hovercraft

Key to the paintings

1 Bow thruster
2 Sound transmitters
3 Observation portholes
4 False nose observation chamber
5 Wooden keel
6 Chain well
7 Cast-iron ballast
8 Water tanks
9 Chef's storeroom
10 Storeroom for engine equipment
11 False nose
12 Wooden stem
13 Forepeak
14 Hatchway to false nose
15 Air shaft
16 Windlass electric panel
17 Shower room
18 Chain stopper
19 Laboratory workbench
20 Fore windlass
21 Mooring light
22 Lower fore hold hatch
23 Storeroom—mechanical equipment
24 Watertight door
25 Upper foredeck hold hatch
26 Air conditioner vent
27 Fore passageway and three cabins
28 Mooring bitts
29 Side-scan sonar winch
30 Crew cabins
31 Mooring lines rack
32 Electrician's workshop
33 Cold water fountain
34 Light alloy mast
35 Radio and radar generators
36 Steering column
37 Ship's bell—"J826 1942"
38 Gyroscopic compass
39 Air conditioner ventilators
40 Helm
41 Masthead light
42 Decca 060 radar and receiver
43 Anemometer wind vane
44 Decca ACS 630 radar transmitter
45 Compass
46 Decca ACS 630 radar receiver
47 Magnavox satellite direction finder recorder
48 Alidade
49 Bridge equipment: Raytheon echosounder, Omega
   receiver, TV, radios, telephone
50 Nikon photo enlarger
51 COMSAT satellite radar dome antenna
52 Captain Cousteau's office
53 Data reader, recorder, distributor
54 NASA flag from Apollo's moon voyage
55 Inter Ocean System computer and data recorder
56 Captain and Mrs. Cousteau's cabin
57 Helm servo motor
58 25kwa 220v alternator and control cabinet
59 Oxygen tanks
60 Magnavox UHF/VHF satellite antenna
61 Electrical distribution switchboard
62 Stern light
63 Fuel oil tanks
64 Four divers' cabin
65 Galley (kitchen)
66 Starboard Junkers high pressure compressor
67 Inflatable liferaft
68 Mess hall
69 Mufflers
70 Compressed air tank
71 Starboard davit
72 Two-man Galeazzi decompression chamber
73 Starboard Allied Water desalination unit
74 Electrical and minor mechanical workshop
75 Ingersoll-Rand air compressor, for diesel starter
76 Hughes 300C helicopter
77 GMC 8-268A eight-cylinder main engines
78 Generator oil tank
79 60kw 110v generator
80 Crane hydraulic winch
81 20kw generator
82 Bergen Nautic hydraulic winch
83 Centrifuge oil filter
84 Helicopter pad
85 Remote-control color television camera
86 Duclos winch control panel
87 Daily gas-oil tank
88 Disposable gas tank
89 CO_2 tanks
90 Duclos winch oil compressor group
91 Starboard propellor shaft
92 Duclos oil tank
93 Hatchway and passageway to aft hold
94 Steel cable drum
95 Line drum for diving saucer
96 Stern hold
97 Watertight compartment
98 Aft peak
99 Aft windlass
100 Rudder control area
101 Helm angle repeater and transmitter
102 Aft windlass engine and reduction gearing
103 Water tanks
104 Stern chain well
105 Diving platform
106 Starboard rudder axle
107 Diving saucer support stand
108 Starboard rudder
109 Propeller shaft bracket
110 Outer propeller shaft
111 Starboard propeller
112 Port propeller
113 Propeller shaft
114 Port rudder
115 Diving saucer support stand
116 Port rudder axle
117 Aft windlass control switches
118 Aft peak
119 Water tanks
120 Aft clamp
121 Yumbo crane hydraulic winch
122 Yumbo crane oil tank
123 Yumbo crane oil compressor
124 Yumbo crane electrical panel
125 Yumbo hydraulic crane
126 Hatch and passageway to aft hold
127 Diving saucer ballast
128 SP-350 two-person diving saucer
129 Duclos electric panel
130 Lubricating oil drums
131 Disposable gas tank
132 Two-drum Duclos winch
133 Fuel oil tank
134 Watertight engine-room door—rear hold
135 Duclos hydraulic winch motor
136 Guinard fire extinguisher pump
137 Interboard telephone
138 Divers' tanks
139 Main lubricating oil reservoir
140 Aquarium for photographing animals
141 Bauer air purification system
142 Helicopter flight deck
143 Main engine bed mounts and oil tanks
144 Turbo-generator starter batteries
145 GMC 8-268A eight-cylinder main engines
146 Engine-room
147 Hatchway to engine room
148 60kw, 110v generator
149 Mechanical workshop

150 Launch
151 Funnel
152 Electrical control transmitter
153 Exhaust pipes
154 Port Allied Water desalination unit
155 Exhaust fan
156 Pressurized fresh water tank
157 Mess hall
158 Compressed air tank
159 Inflatable lifeboat
160 Fuel-oil tanks
161 Port Junkers high pressure compressor
162 Flagstaff
163 Stern light
164 Magnavox satellite antenna
165 Captain's cabin
166 Galley (kitchen)
167 Factory area storage cabinet
168 Oxygen tanks
169 Storage room for Johnson outboard motors
170 Captain and Mrs. Cousteau's cabin
171 Divers' radio receivers and transmitters
172 Photocopier
173 Refrigerant compressor
174 Captain Cousteau's office
175 EGG side-scan sonar receiver/recorder
176 EDO wave guide and transceiver
177 Graphic recorder for EPC sounder
178 Batteries
179 Radar aerial for Decca ACS 630-44
180 Loudspeaker
181 Alidade
182 Red and white hydrographic lights
183 Radar aerial for Decca 060-42
184 Cousteau Society pennant
185 Bridge equipment—Raytheon echo sounder, Omega receiver, TV, Ben Galathee log, radios, telephone
186 Masthead light
187 Forecastle searchlight
188 Scott signal searchlight
189 Main engine remote-control device
190 Compass
191 Chadburn
192 Downpipe for radar wave guide
193 Revolving windshield

194 Helm
195 Ship's bell
196 Light alloy mast
197 Boom for lowering side-scan sonar into the water
198 Air conditioning ventilators
199 Radio and radar generators
200 Winch for lowering side-scan sonar into the water
201 Crew cabin
202 Side-scan sonar transmitter
203 General staff and passenger cabin
204 Fore hold hatch
205 Cold storage rooms
206 Fore windlass
207 Wine cellar
208 Toilet
209 Hatchway to false nose
210 Shower
211 Electric water heater
212 Storeroom for expedition equipment
213 Hawsers
214 Freezer
215 Storeroom
216 Water tanks
217 Cast-iron ballast
218 Chain well
219 Keel
220 False nose observation chamber
221 Observation portholes
222 Sonic depth-finder transmitters
223 Bow thruster
224 Red and green signal lights
225 Captain and Mrs. Cousteau's sink, shower, and toilet
226 COMSAT transmitter/receiver switches
227 4kwa commutator
228 Lathe and drill
229 Zodiac
230 Junker intake tube
231 Engine ventilator
232 Diesel starter air tank
233 Cameraman's cabin
234 Electric pump
235 Stabilizing keel
236 Chart table
237 Saltwater pump
238 Return pulleys for rudder rods

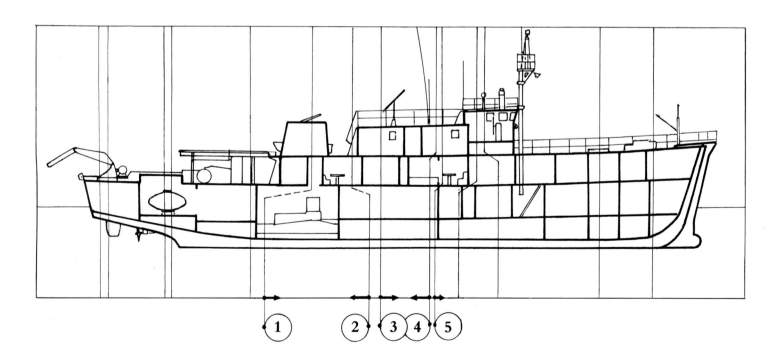

Following are five cross-sectional paintings of *Calypso*'s interior; the views are either from fore or aft, as indicated in the diagram above.

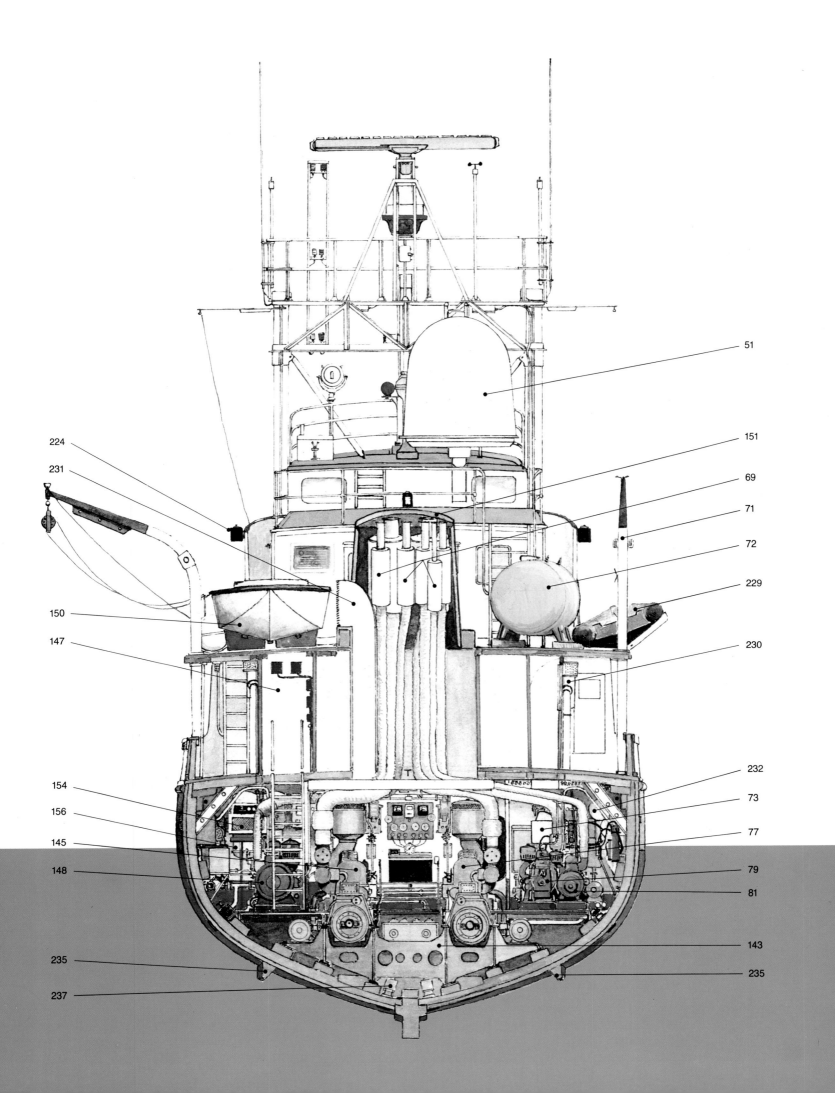

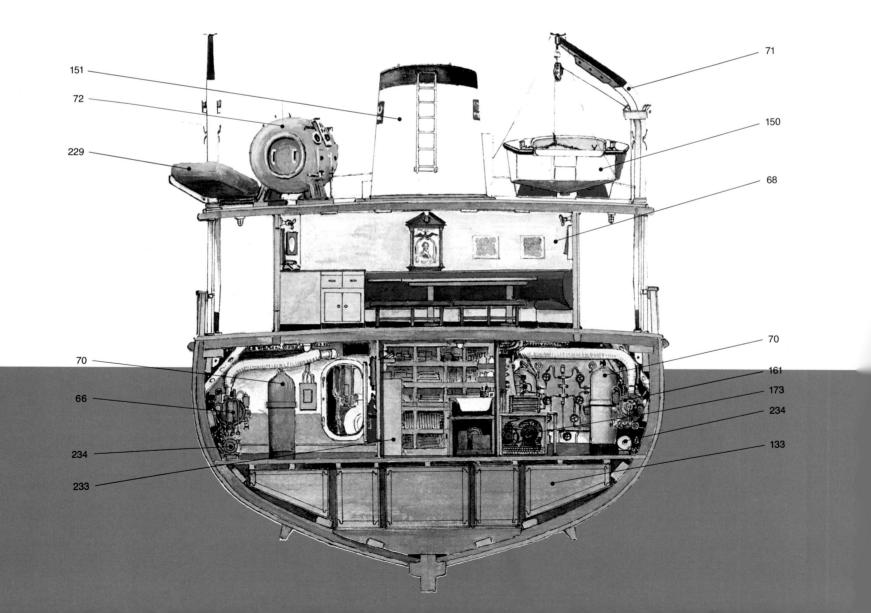

151

72

229

71

150

68

70

70

66

161

173

234

234

233

133

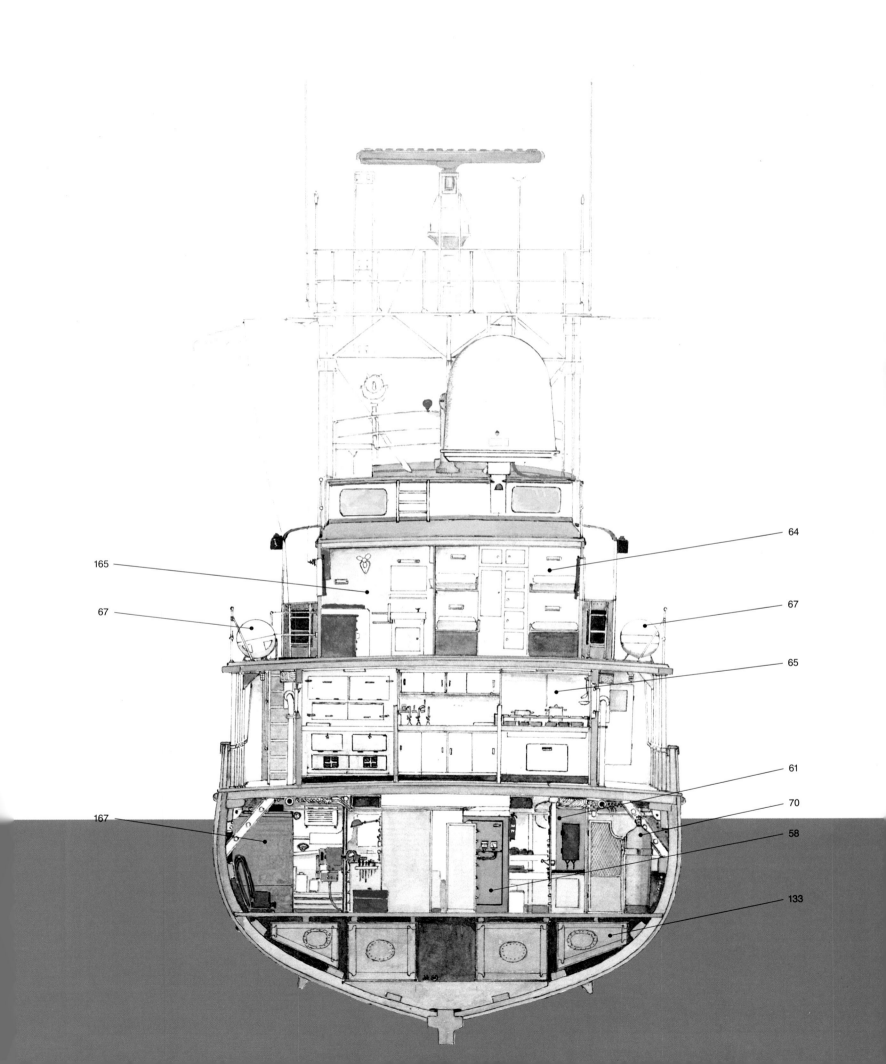

64

165

67

67

65

61

70

167

58

133

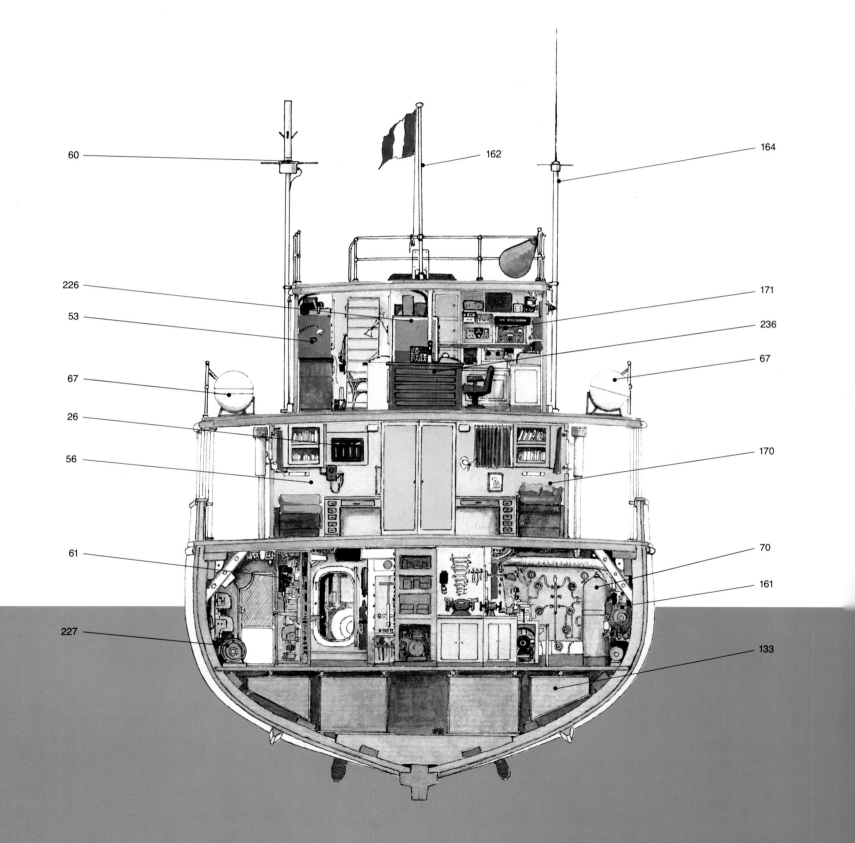

60

162

164

226

171

53

236

67

67

26

170

56

61

70

227

161

133

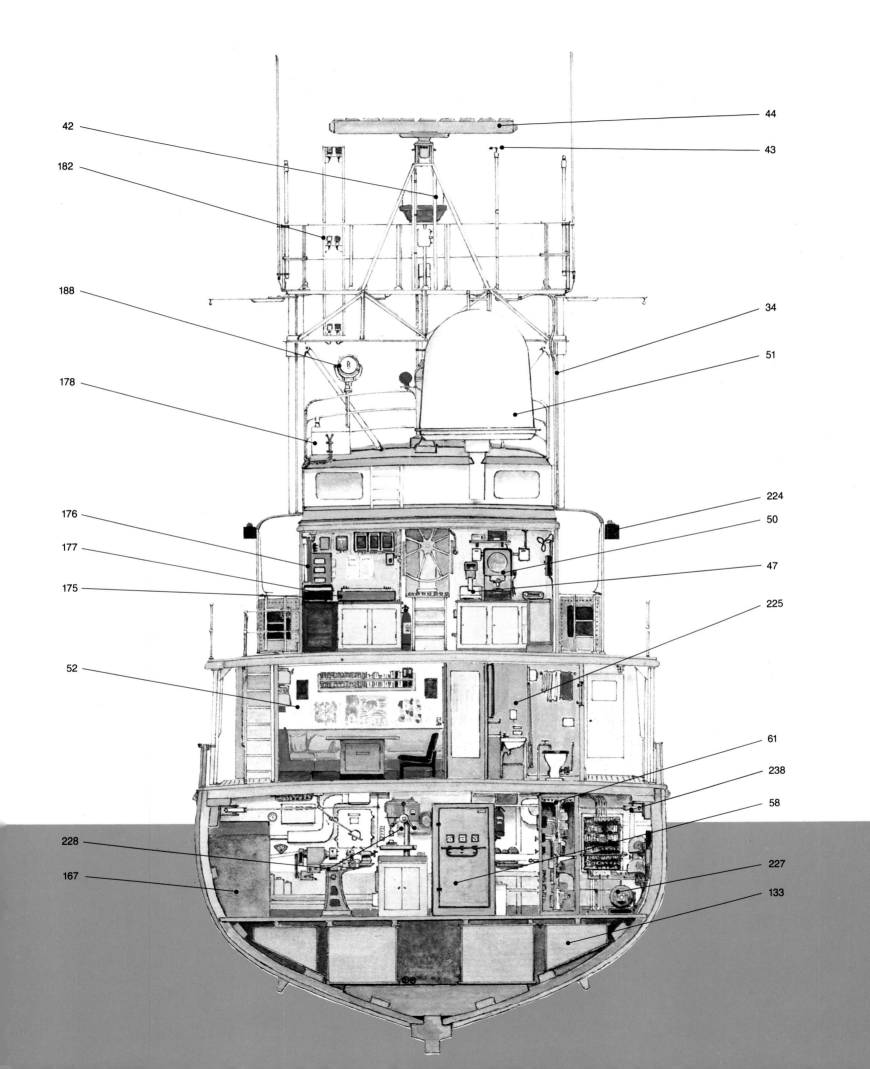

44

42

182

43

188

34

51

178

176

224

177

50

175

47

52

225

61

238

58

228

167

227

133

SOUNDING
THE DEPTHS

The winter of 1955–56 saw a lot of maintenance work done on *Calypso*, much of it by the crew themselves. On March 19, 1956, a new electrician joined *Calypso*: Jacques Roux, from then on called simply Gaston.

During the early spring, *Calypso*, under the direction of Professor Bourcart, was devoted to the taking of geological samples by the use of corers and dredgers. Work was concentrated off the coasts of Villefranche and Nice, and off Ajaccio, Porto, Cargèse, and Ile-Rousse, all in Corsica.

On April 28, *Calypso* and her crew spent two days at Cannes, where the film *The Silent World* was being shown at the Cannes Film Festival. To the delight of all, the documentary was awarded the Palme d'Or, the highest award given at the festival. Later, the film would win for Cousteau his first Academy Award.

On May 2, after being joined aboard by Jacques Forest of the National Museum of Natural History and a team of scientists, *Calypso* set off on a new expedition—this time to the west coast of Africa.

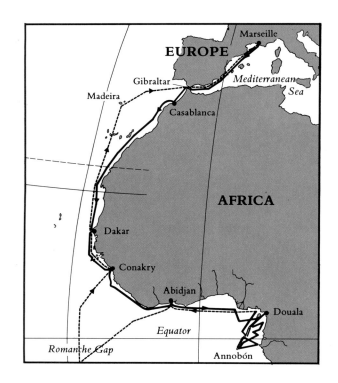

The first part of the expedition was to concentrate on the study of the marine environment and fauna off the coasts of Guinea, Cameroon, Senegal, and the Ivory Coast. Special attention was to be given to the Portuguese islands, sometimes called the Chocolate Islands, in the Gulf of Guinea. Under the command of Captain François Saout, *Calypso* moved south and followed the Spanish coast in rough seas, arriving in Casablanca on May 6. While traveling from Casablanca to Dakar, the crew fished for tuna and ran into many colonies of jellyfish. On May 11, *Calypso* reached Dakar, in rather bad seas. From there ports of call gave way in rapid succession to one another—Conakry, Abidjan, Douala—for Forest had set up a schedule for *Calypso* that required the crew to work night and day. They made hundreds of stops at specified points en route to measure, observe, and collect samples for various hydrological and biological studies. Nets, dredgers, trawls, bottles, thermometers were all forever swinging overboard in the pursuit of this information.

From the second of June to the sixteenth of July the activity was centered around the coast between Douala and Gabon's Port Gentil, as well as around the islands of Fernando Póo, Príncipe, São Tomé, and Annobón. The crew encountered on the Annobón Islands an epidemic of sleeping sickness spread by the tsetse fly, and discovered that the Portuguese soldiers had been treating the illness in their own manner: by shooting all mammals on the island—cats, dogs, cows, pigs.

On June 25 *Calypso* set sail for the Romanche Gap—the deepest trench in the Atlantic Ocean—situated about one-third of the way from Africa to South America. On board, in addition to Captain Cousteau and the usual crew members, were Dr. Edgerton and André Laban, a diver (and underwater painter). Laban had with him an innovative and unique implement: a special grooved wheel that measured five feet in diameter and weighed 440 pounds. This wheel would be instrumental in the success of *Calypso*'s next experiment—a revolutionary anchorage at a depth of over 24,000 feet.

Until this mission, anchorages at depths greater than those permitted by standard chains had been attempted only by means of an anchor and a steel cable whose weight increased in proportion to its length. The result was that the breaking point of the cable was often reached before the anchor touched bottom. Several alternatives had been tried to overcome this deficiency, but were largely unsuccessful. Special, expensive steel cables of progressive diameter were employed with limited success, as were nylon lines, which fared little better. Although nylon was a likely option, because it weighed practically nothing in the water yet kept its strength no matter what its length, it also had a fatal drawback: its elasticity. When wound under tension, it would put such a severe strain on the winch's drum that it would cause the drum to shatter.

The solution adopted aboard *Calypso* was to haul up the nylon cable by passing it twice around Laban's famous "wheel," which was bolted on the axis of the main power winch. The cable could then be wound, with virtually no tension, on a separate drum. (With all the gear necessary for a record-depth anchorage, *Calypso*'s quarterdeck was transformed into a spiderweb.)

After three days' sailing, *Calypso* reached the equator, eight hundred nautical miles off the coast of Africa. She was now nearly in position perpendicular to the Romanche Gap, only two or three miles wide but 24,928 feet deep.

When it actually came time to put Laban's wheel into use at such a deep anchorage, several possible ramifications had to be considered. For example, although the advantage of nylon was its light weight in the water, this very characteristic was disadvantageous when it came to making an anchorage. Contrary to heavy chains, which pull horizontally on the anchor when a ship strains, nylon tends to pull toward the upper part of the anchor's shank, causing the anchor to drift. Cousteau devised a solution to the problem based on the wooden anchors of antiquity; these were linked to their ships by simple ropes made of hemp or jute, which, like nylon, pulled the anchors upward. The wooden anchors of the past, however, were heavily ballasted by lead anchor stocks, which counteracted the upward-pulling tendency.

Following this principle, *Calypso*'s anchorage was made up of a 700-pound anchor, 100 feet of heavy chain, a large pig iron weighing 550 pounds, a 200-foot steel cable, and the thin line of braided nylon, less than a quarter-inch in diameter and 32,800 feet long.

It took two and a half hours to let the anchor line out the first time, for it was a delicate matter to ensure that it would not get tangled on the way down. The operation was a success, however, and the anchorage was deeper than any that had been accomplished before. The echo sounder showed that *Calypso* was not dragging, despite a moderate current.

Professor Edgerton took advantage of the ship's steadiness by lowering his photographic equipment on another nylon line. The camera and its electronic flash provided remarkable pictures of marine life heretofore unseen at such great depths. A detailed map of the trench bottom was obtained by transferring the anchor line to a launch equipped with a radar beacon. The map proved the Romanche Gap to be part of the famous rift valley of the central Atlantic, the fault area where the forces of the earth's continent-sized crustal plates come together. Meanwhile, *Calypso* made transverse soundings with reference points carefully marked in relation to the launch. These soundings enabled Christian Carpine and André Laban to update their maps and to make a plaster mock-up of the trench.

Calypso's anchorage was interrupted on the second day, when tragic news reached the ship. Professor Harold Edgerton's eldest son had accidentally died while testing a new closed-circuit oxygen diving device in New England. Everyone was deeply saddened, the more so because of the deep affection felt for "Papa Flash." The anchorage at the Romanche Gap was immediately abandoned, with the five miles of line, which would have taken too long to bring up, sacrificed. A fast course was steered toward Conakry, where Edgerton and Cousteau caught a flight to the United States. Plans were made for Cousteau to rejoin *Calypso* in Madeira.

Calypso soon sailed again, first to Dakar, and then to Funchal in Maderia, where Edgerton's underwater photographic equipment was again tested in both deep and shallow waters. From there, *Calypso* returned to Marseille for a two-week layover.

On September 10 *Calypso* started on a program to study, under the direction of Professor Yves La Grande, the optical qualities of the waters off the coast of Corsica. The penetration of light in deep water, its absorption and diffusion, and the percentage of particles at different depths were measured at such places as Girolata, Gargalo, Ajaccio, and Calvi before *Calypso* returned to Monaco on September 17.

On September 21 *Calypso* left on a new hydrology mission, with Henri Lacombe. Battling fierce winds and dreadful weather the entire way, she sailed from Marseille to the Ionian Sea, where the survey was begun. Off the island of Crete, the clutch of the starboard engine broke down, and *Calypso* was forced to take shelter in a nearby cove while the crew attempted to make repairs. Professor Lacombe's ill-fated program was finally completed on October 23, when Professor Jean Marie Pérès came on board to begin biological studies. No sooner had *Calypso* left port, however, than Captain Saout received orders to return immediately to Marseille. War between Israel and Egypt had just broken out.

Professor Pérès was greatly disappointed, for he knew the po-

Professor Edgerton and Captain Cousteau.

tential value of his biological expedition. He had planned for *Calypso* to sail between the Greek and Turkish islands, but these countries too were experiencing serious diplomatic tensions. *Calypso* dropped off one of Pérès's staff, a Greek citizen, at Samos, then left for home as quickly as possible. She arrived safely at Marseille on November 10, although she encountered many warships in the Aegean Sea. After a short mission to the coast of Genoa, she returned to her home port of Toulon for winter respite on the ninth of December.

Following the directives of Captain Cousteau, *Calypso* spent her 1957 winter layover in the Antibes shipyard, where she underwent substantial alterations that were to change her appearance considerably. The deckhouse, smokestack, and catheads (those parts of the ship to which the anchor line is fastened) were all dismantled, and the spar deck was lengthened, the diving area redone, the ventilation system for the engines modified, and the area of the chart room tripled in size.

On March 30 *Calypso* left Antibes for Marseille, where a larger hold panel and a false funnel were installed. The instruments were moved into the newly designed chart room, which had seemed so spacious on paper but quickly proved to be cluttered by the profusion of navigation devices.

Once she was completely outfitted, *Calypso* was ready to depart on her latest expedition. On April 23 she sailed to Villefranche, where, together with the *Elie Monnier*, Cousteau's old charge, and the bathyscaphe FNRS III, she collaborated on a series of test dives. Grégoire Trégouboff, director of the zoological laboratory at Villefrance-sur-Mer, joined up with *Calypso* to lead a program of study of the migration of plankton.

Calypso's next mission, directed by Jean-Paul Brouardel, took her once more to the west coast of Corsica, where her sensitive instruments measured the amount of oxygen at the bottom of the sea.

On June 1 *Calypso* was unexpectedly called back to Marseille, where she was to participate in the first Eurovision program organized in France by Radio-Télévision Française. The broadcast, called "Live from the Sea Floor," was to document the activities of Cousteau and his *Calypso* team at some of their typical work sites. Three main locations were chosen for filming: the ancient shipwreck off the coast of Grand Congloué; a modern sunken ship off the island of Frioul, near Marseille's Vieux Port; and the port of Marseille. Celebrated contemporary commentators, such as Igor Barrère, Pierre Tchernia, Georges de Caunes, Roger Gouderc, and Alexandre Tarta, enlivened the program. "Live from the Sea Floor," the first program of its kind in the world, was televised on June 15 to great critical and popular acclaim.

Immediately after the broadcast, *Calypso* resumed her scientific functions. France's participation in the International Geophysical

ASSIGNMENT IGY

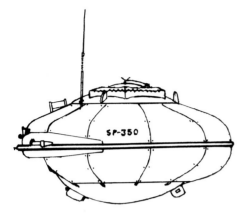

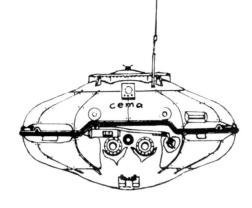

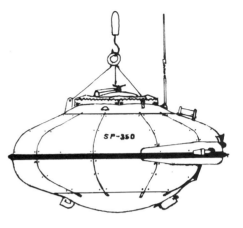

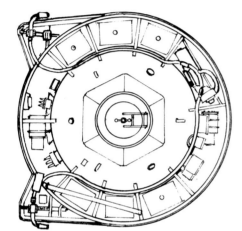

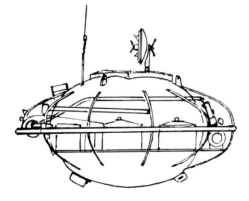

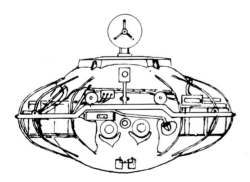

Year included plans to study the character and distribution of the Mediterranean waters that move westward along the Strait of Gibraltar and flow into the Atlantic Ocean, where they slowly mix with the general circulation. Much was known already about the unusual system of currents in the Mediterranean: waters from the Atlantic Ocean entering the Mediterranean create a powerful surface current running east, while a smaller amount of Mediterranean water moving down along the Gibraltar shelf creates a deep current running west. Even such early seafaring peoples as the Phoenicians had managed to devise ways to overcome the force of the currents, but never had the reasons behind them been understood. How were these waters distributed? To what extent did they flow? And how were the differences in temperature maintained? These were just some of the questions that *Calypso*'s crew, under the direction of Professor Lacombe, sought to answer.

Revolutionary new instruments to measure and record the currents and swells of the area were soon brought aboard, as were water bottles specially designed to be lowered to specific depths of the sea in order to bring back samples of water that would then be tested for, and classified according to, temperature, salinity, and other factors. Much useful information was gathered.

On July 12 Captain Jean Alinat took temporary command of *Calypso* from Captain Cousteau and sailed for the Atlantic Ocean, passing Gibraltar on the sixteenth and arriving at Brest on the twentieth of July. From Brest *Calypso* began a three-month program journeying in the service of France, fulfilling that country's pledge to science in honor of the International Geophysical Year. Onward she sailed, often in heavy seas and inclement weather, stopping at many ports along the western coast of Europe to explore the mysteries of the currents.

Once this program was completed, *Calypso* returned to Marseille, where a new challenge awaited her. She arrived on October 5 to test the first full-sized prototype of Cousteau's projected diving

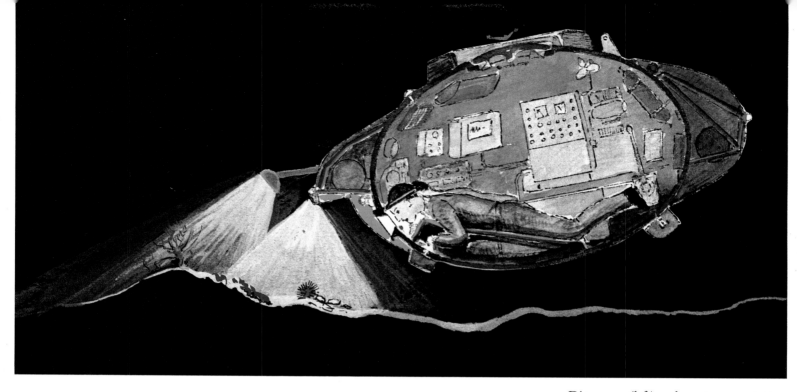

saucer, the hull of which had been completed by the OFRS during the summer months. On October 25 the saucer's hull was hoisted aboard, and *Calypso* set off for the island of Riou, near Grand Congloué, where, under the actual conditions it would face in use, the hull was to be tested for durability, maneuverability, and the ability to withstand the pressure unique to the depths of the sea. The saucer was heavily ballasted, attached to a steel cable with a strong braided-nylon line, and lowered into the waters off the shores of Riou. The results of all the tests were satisfactory, and the OFRS was instructed to proceed with design along these lines.

The diving saucer itself was not the only thing of concern to the Office Français de Recherches Sous-marines that fall. Aware that the finished saucer would weigh over four tons, Captain Cousteau had long since charged the OFRS to provide or develop some piece of equipment that would be able to easily maneuver the heavy saucer into and out of rough water with no danger to it or to *Calypso*'s crew. Cousteau had in mind something on the order of a special boom he had seen employed by the British in the Persian Gulf some three years before, a boom that was able to place a sensitive object like a gravimeter in the water and recover it without allowing it to swing from the end of the cable. The instrument developed by the OFRS must also clamp on to the saucer before it is removed from the water, to ensure that the vehicle would not get lost in the sea.

Miraculously, such a piece of equipment, already existing, was discovered in September at the annual Marseille Fair. It was a hydraulic crane with a five-ton capacity that had been used for public works but seemed suitable for adaptation to maritime use. The construction engineer of the Fair agreed to help adapt the crane, nicknamed "Yumbo," for use aboard *Calypso*. It would soon help to revolutionize the maneuvering of heavy equipment aboard ships, and serve as a model for many other seagoing vessels.

After modification to both Yumbo and *Calypso* herself (her wooden deck, for example, was reinforced to enable it to support

Yumbo's great weight), the crane was delivered and installed in early January 1958. The crane was positioned aft on *Calypso*'s port side, so that it was able to lift the saucer out of the sea from either the rear or the side of the ship.

By March the diving saucer had been fitted with observation portholes and strain gauges to measure the stress on various parts of the hull, and was ready to be tested further. The saucer, now loaded with one and a half tons of ballast, was lowered from *Calypso* on the nineteenth of the month, down 2,000 feet into the depths of the sea. This in itself was no easy task, as the lowering and raising of the saucer both required considerable care and attention. While attempting to bring the saucer back aboard *Calypso* from the sea, the boatswain made sure to maneuver the winch cautiously, for he knew that jerking the cable too violently could have disastrous effects. The ascent therefore went smoothly, until suddenly, just as the saucer approached the surface of the sea, *Calypso* veered sharply. The cable snapped, lashing about the quarterdeck like a giant whip, and sent the diving saucer plunging back into the sea.

Luckily, no one was injured seriously in this accident; indeed, only one person, Raymond Coll, was hurt at all, and he escaped with a superficial cut on his cheek. But the diving saucer, separated from the crane and its cable, was nowhere to be seen.

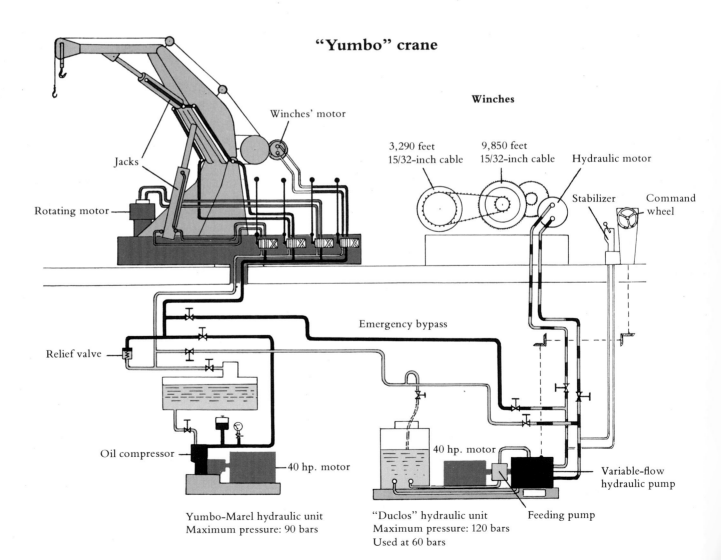

"Yumbo" crane

Winches' motor

Winches

3,290 feet 15/32-inch cable

9,850 feet 15/32-inch cable

Hydraulic motor

Jacks

Stabilizer

Command wheel

Rotating motor

Emergency bypass

Relief valve

Oil compressor

40 hp. motor

40 hp. motor

Variable-flow hydraulic pump

Feeding pump

Yumbo-Marel hydraulic unit
Maximum pressure: 90 bars

"Duclos" hydraulic unit
Maximum pressure: 120 bars
Used at 60 bars

Albert Falco operates Yumbo to swing the troika camera sled over the side.

Calypso returned to the site the next day to try to bring the saucer back aboard. By means of the highly accurate echo sounder, the crew found that the diving saucer was floating, 33 feet above its ballast, at a depth of 3,280 feet. It was evident that the portholes and hull were not damaged even at the great depth—which indicated that the saucer had withstood the pressure even better than had been hoped. But there was no way to reach the saucer at that depth, and so the salvage operation was suspended. *Calypso* returned home, leaving the first diving saucer over half a mile beneath the surface of the Mediterranean.

Calypso's next expedition entailed not one but a dozen new research programs, involving hydrology, biology, seismology, and geology in various studies off the coasts of Marseille, Nice, and Corsica. While he was between two of these missions, Cousteau, examining precision sonar recordings, noticed at a depth of 8,500 feet some remarkable hills, which he later interpreted as salt domes. These salt domes, beacons for possible oil deposits in the deep waters of the Mediterranean, would figure prominently in some of *Calypso*'s future missions.

On June 26 *Calypso* was loaded with new technological equipment, took aboard naturalists from Paris, Marseille, and Monaco, and went off on a three-month hydrological-biological program. Madeira, Lisbon, and Porto in Portugal, Barcelona and the Spanish island of Alborán in the Mediterranean, and Tangier and Spanish Ceuta on the Moroccan coast were just a few of the many ports of call on this long sea-faring voyage. While dredging at Alborán, the crew encountered a monstrous species of kelp, its stem more than six and a half feet long and its leaves more than a foot wide. Falco, diving down, soon discovered a veritable forest of these kelp at depths between 130 and 140 feet. The time soon came when *Calypso* had to move on, but not before it was determined that she should return soon to that site to further investigate this remarkable undersea forest.

On September 23 *Calypso* was back at her home port to test a new invention designed by Jacques Cousteau and developed by the OFRS. This device, a photographic sledge suitable for towing along the bottom of the sea, was dubbed the "troika" because of its resemblence to a Russian sled, with its top arch and two runners.

The idea for such an instrument first came to Cousteau when he was diving in a bathyscaphe, a large deep-sea diving sphere, at a depth of around 5,000 feet. Cousteau was fascinated by the bathyscaphe's potential for deep-water exploration, but he was also aware of some of its drawbacks. For one thing, the bathyscaphe was extremely expensive to operate; for another, it was much too heavy and cumbersome a vehicle to be carried aboard *Calypso*. Still he realized the need to be able to obtain photographs and motion picture films from the greatest depths. From his musings sprang the idea for the troika, an unmanned device able to be towed at depths of more than 25,000 feet and equipped with automatic Edgerton cameras and flash guns. It was designed so that even if it tipped over while being towed, the cable's traction would act to right it.

The troika soon proved to be an invaluable contribution to the oceanographic community. In years to come, troikas would take many tens of thousands of previously unattainable photographs and films in very deep waters.

Once the prototype troika had been tested, *Calypso* returned, on October 4, to the island of Alborán for more extensive diving amid the large kelp forest discovered by Falco in August. The kelp was determined to be laminaria, which grew to a height of from 20 to 26 feet. The forest extended for over half a mile, and looked to the divers somewhat like a plantation of banana trees. This was indeed a magnificent site, with large silver leerfish circling in schools above the forest while scorpion fish lurked below the vegetation. A film was shot there to record the striking beauty of this lush undersea world.

Calypso sailed from the kelp forest of Alborán to Nice, Cannes, and Genoa, where her scientific and diving teams undertook, with

A crewman mounts one of the troika cameras.

Professor Trégouboff leading, a new study of plankton in the area. Then back she went to her home port of Toulon for her annual winter maintenance.

Beginning in January 1959, frequent discussions were held aboard *Calypso* with experts and technicians from Gaz de France, the French National Gas Company. Officials of the company had proposed to Captain Cousteau a new project: a thorough study of the possibility of laying natural gas pipelines from Algeria to Spain at depths of over 7,000 feet. These pipelines would be used to supply natural gas from sources in North Africa to most of the European continent. The task aboard *Calypso* would be to organize and carry out the broadest, most comprehensive, and most accurate study done up to that time. The results of this important work, which would use much deep-water photography, would determine if and where the pipelines between North Africa and Spain could be submerged.

Marine charts alone could not be relied on for sufficient information for this delicate venture. Unlike transoceanic cables, which can be submerged almost at random, the laying of the steel pipe needed to transport natural gas required extensive knowledge of even the slightest irregularities of the bottom of the sea. It was also necessary to know corrosion factors, such as the presence of dissolved oxygen or anaerobic bacteria, and the mechanical resistance of the soil.

It was decided that the first step would be to verify the reliability of the studies already made on the shortest route between the two continents—that is, across the Strait of Gibraltar. On February 5 *Calypso* arrived in Tangier with Captain Saout in command and Captain Alinat and several technicians from Gaz de France also on board. It took a week of arduous underwater and onboard activities to make all the necessary soundings. They then sailed for Marseille to analyze the results of these soundings. They concluded that the route across the Strait of Gibraltar was not at all a desirable one. Plans were made to study another route, this one between Mostaganem in Algeria and Cartagena in Spain—a route that spanned 112 miles of open water at depths reaching 7,870 feet.

Making accurate soundings required the use of the most up-to-date technology available. *Calypso*'s sonar system was capable of measuring depths precisely to within a foot and obtaining information on the consistency of the sediment as far as ten feet below the sea floor, but this information is only useful if the ship's position can be pinpointed within a few yards. The network selected for this mission was a "Decca two-range" system, in which the main transmitting station was installed aboard *Calypso* while the secondary transceivers, called "slaves," were set up ashore on both sides of the Mediterranean—a green one at Aguilas, near Cartagena, and a red one at Cape Falcon in Algeria. A forty-five-foot antenna was erected on the ship's quarterdeck and secured on all sides by braces

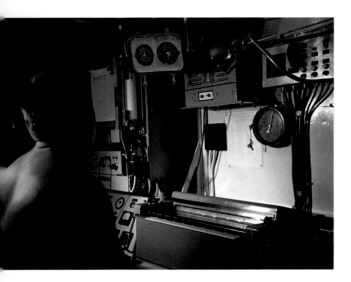

A *Calypso* technician monitors the deep sonar depth recorder to help determine the safest path for a gas pipeline linking Algeria and Spain.

crisscrossed with insulators. This system measured *Calypso*'s positions accurately within 33 feet during the daytime and 98 feet at night, distances that were judged to be sufficiently accurate to chart every sea-floor obstacle. A plotter installed on the bridge automatically traced the course followed by *Calypso*.

Calypso left Marseille for Cartagena on March 24 with Captain Alinat as skipper, for Captain Cousteau was occupied with administrative problems ashore and could only come aboard at certain intervals. The ship made endless shuttles between Spain and Algeria, during which, in addition to mapping the area, the crew dredged, took core and bottle samples from the sea, and drew bacterial samples from the mud. Also, and perhaps most importantly, *Calypso* towed behind her for hundreds of miles at great depths the troika, which took thousands of photographs of the marine environment.

Three months after the work had been started, the final fixes were taken, and, after celebrating with champagne, the team returned to Marseille. The data, analyzed in Monaco, Paris, and Marseille, showed that there was indeed a channel safe for the laying of the gas pipelines between Cartagena and Mostaganem. Ironically, however, Algeria decided for political reasons to transport the gas in liquid form instead of through pipelines. The expedition was not a loss, though, for it provided *Calypso*'s team with the opportunity to put into actual practice for the first time such innovative equipment as the troika and the Decca two-range plotting system, not to mention the additional knowledge they gained about the Mediterranean waters and sea floor.

On July 25, 1959, the *Calypso* team finally got to launch the world's first practical minisub, or diving saucer. The two-man exploratory submarine was formally christened the "*soucoupe plongeante*" (diving saucer), or SP-350, and was equipped with a cinematographic camera, a still camera, and an external hydraulic lift and pincer and storage basket for the collection of bottom specimens. Tests proved the SP-350 to be perfectly watertight and easily maneuverable. The initial dive, which took place off Marseille in the presence of the Secretary for Merchant Marine, with Jean Mollard and Albert Falco at the controls, was the first in a long series of similar dives. From the moment of its initiation, this revolutionary minisub became one of *Calypso*'s most indispensable tools. It completely changed the character of underwater exploration, affording for the first time man's active presence in waters as deep as 1,150 feet.

WELCOME TO NEW YORK

Calypso sailed from Marseille on July 29; her eventual destination was New York, where she was due on September 1, 1959, for the first World Oceanographic Congress. Never one to waste precious sailing time, Cousteau organized an expedition for the occasion: the exploration of the flat tops and slopes of several seamounts, or guyots, and the inner walls and bottom of the famous Atlantic Rift Valley. The expedition was carried out with a team of biologists from Marseille and was under the direction of American geologist Lloyd Breslau.

The seamounts to be explored were truncated volcanic cones rising up some 15,000 feet from the abyssal ocean floor. The exploration of these features was only made possible by the development of the troika as a mobile, deep-diving photographic platform.

The use of troikas in this instance was in itself a delicate operation, for they could not safely be towed along the bottom of the rocky and mountainous terrain at a speed greater than two knots. As usual, however, *Calypso*'s team had devised a solution—a way to ensure the maintenance of such a slow speed and thus to minimize the chance of damage to the troikas. They employed a revolutionary new kind of stationary reference point by using "kytoons"—streamlined helium-filled balloons similar to the "sausage" balloons of World War I, though much smaller. Positioned on the shallow tops of the seamounts and across the middle of the Atlantic Rift Valley, the kytoons had been inflated aboard *Calypso* and then anchored by thin but strong nylon lines to grapnels of about forty-four pounds. Each one carried a light aluminum-foil radar target. By using the radar aboard *Calypso*, the navigator could take his bearings from the kytoons and therefore adjust the ship's speed and course relative to the bottom of the ocean, despite the force of the current.

Even with these precautions, one of *Calypso*'s troikas was lost during the course of the program, for the deep-lying rocks of the

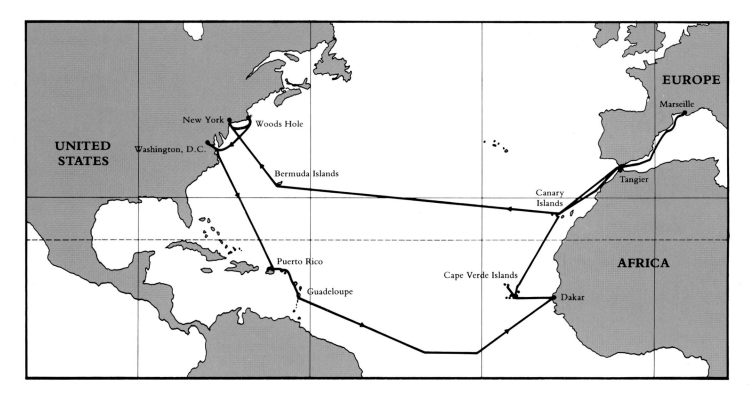

area were very treacherous indeed. Still, since the troikas took 660 pairs of color stereo photographs during each operation, *Calypso*'s team were able to secure many thousands of pictures vital to increased knowledge of this area.

Once exploration of the Atlantic Rift Valley had been completed, *Calypso* resumed her voyage to join the World Oceanographic Congress. After a brief stopover in Bermuda, she arrived in New York on August 29, 1959, the eve of the Congress, to a spectacular reception. Her triumphal welcome to America was on the same scale as that received by such celebrated ships as the *Normandie* or the huge ocean liner the *Queen Mary* on their maiden voyages. An escort of hooting tugboats, waterworks from the city's fireboats, and ear-piercing sirens from both ships and docks saw

Falco, Maurice, and Bonnici prepare to release a kytoon—a helium-filled balloon bearing an aluminum-foil radar target, that will provide a stationary reference point for mapping the sea floor.

Calypso into New York harbor; a dirigible hovered overhead taking photographs. All were on hand to receive the little former minesweeper rescued from naval surplus by Cousteau.

While berthed in New York, *Calypso* was visited by thousands of scientists, city officials, navy representatives, journalists, and photographers, all of whom knew of her unique role in oceanography. Meanwhile, Captain Cousteau was busy at the Congress showing the extraordinary photographs of and commenting on the rich statistical information gathered about the strange undersea mountain range in the mid-Atlantic.

After the World Oceanographic Congress, *Calypso* did not leave the United States immediately. Rather, she first sailed to the renowned Oceanographic Institute at Woods Hole, Massachusetts, then traveled south and then up the Potomac River to Washington, D.C., where she was warmly received as a guest of the National Geographic Society. She then set off for her native France in mid-September, at a leisurely pace, stopping variously along the way to carry out some impromptu scientific missions.

September is often the month of tropical storms in the Atlantic, but it is also a time of calm periods favorable to the many difficult operations *Calypso* was to undertake at sea. The crew decided to take advantage of one calm stretch by photographing the deepest point in the Atlantic Ocean: the Puerto Rico Trench, 26,240 feet—nearly five miles—below the surface. Again this was to be accomplished with the help of a troika equipped with Edgerton cameras and flashes and towed along the bottom.

Because of the great depth of the trench, it was necessary to tow the troika on an especially powerful winch with an enormous drum wound with 40,000 feet of cable of progressively greater strength. The ship had to move at a speed no greater than one mile per hour; again, because of varying subsurface currents the navigators could not rely on surface speed to control the ship's position, so the kytoons were once more called into use. Cousteau attempted to anchor one of the kytoons on a thin nylon line with a buoy at the surface, but three times the rope was severed by a shark, and the use of the kytoons had to be abandoned. However, with some acrobatic maneuvering, *Calypso*'s team managed to get several series of six hundred photographs of the bottom of the trench, which revealed for the first time numerous traces of life even at such great depths.

Soon after examining the Puerto Rico Trench, the members of the expedition turned to further testing of the diving saucer, first near the western coast of Puerto Rico, then, under less than optimum weather conditions, at Guadeloupe, off Pigeon Island, in the Caribbean Sea. Unfortunately, the nickel–cadmium batteries used to energize the saucer were not properly insulated, and an underwater explosion, which occurred while Albert Falco was piloting the saucer and Harold Edgerton was aboard as an observer, put the trials to an immediate stop.

On her way home, *Calypso* continued to take soundings and photographs with the troika, stopping often and at various ports to gather additional scientific data. On November 11 she pulled into Dakar to pick up Dr. Jacques Forest, who embarked to lead an important study of the marine fauna of the Cape Verde Islands. While at Cape Verde, *Calypso* received by air a new shipment of nickel–cadmium power cells, and diving in the saucer was resumed. But another explosion soon occurred, this one with Falco and Captain Cousteau aboard, and the two were forced to jettison the emergency lead weight and return to the surface. Further testing of the saucer was again temporarily abandoned, for it was determined to be too dangerous. Cousteau later decided to use the more conventional battery cells instead of the nickel–cadmium ones, however, and the saucer was used from then on without any further explosions.

On Christmas day, 1959, *Calypso* finally returned to her home port of Toulon, where she remained docked until January 22, 1960. Now began her second decade of undersea explorations.

Overleaf: passing into New York Harbor, *Calypso* receives a splashy welcome from a city fire boat.

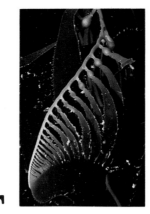

The year 1960 was one of transition for *Calypso*, marked first by the retirement, after nine years aboard *Calypso*, of Captain François Saout. He was replaced by Captain Alinat as interim skipper for the first expeditions of the new year.

Experiments to determine the capabilities of the diving saucer continued to occupy *Calypso*'s scientific and diving teams. The first mission of 1960, led by Professor Pérès, took them to Calvi and Ajaccio on the western coast of Corsica, to study in detail the structure of the steep canyons that dent the continental shelf there. The diving saucer was used several times for this purpose, with Albert Falco at the controls. His first passenger was Captain Cousteau, the second Professor Pérès. During these dives, they observed several of the octopuses that lived in the corners of curious, semi-rectangular enclosures, the walls of which they had built themselves using stones, pieces of pottery, or debris found in the vicinity.

On April 11 Roger Maritano joined *Calypso* as her new captain. Also new to the team was Dr. Romanovsky, who conducted many scientific programs in the next months. One of these studies used flow meters to measure under-surface currents and photographic troikas in an attempt to find correlations between the velocity of the currents and ripple marks on the bottom sands at Villefranche, Corsica, and off the coast of Elba. With Professor Trégouboff the teams also made examinations of fish and plankton in the area and with Professor Bourcart they took corings and soundings and produced seismic profiles with the help of two new high-intensity sound transmitters, nicknamed "thumper" and "boomer," that were invented by Dr. Edgerton.

Calypso left Corsica in May, this time bound for Greece and a program directed by Professor Pérès. Standard hydrological studies and soundings were made along the Euboea Channel, off Volos, in the Gulf of Corinth, and in many other Greek waterways in collaboration with several Greek scientists; and dredgings and photographic profiles of the Matapan Trench were made with the invaluable troika, an instrument still unique to *Calypso*.

The summer months were occupied with many other innovative programs in the Mediterranean: hydrology off the coast of Algeria with Professor Ivanoff; studies of radioactivity in the western Mediterranean; collecting fish and plankton specimens off the

Plans for a New Decade

Top: photographed off the coast of California, this kelp—of the species *Macrocystis*—can grow at a rate of nearly 18 inches a day and reach 200 feet in length at maturity, making it the fastest growing plant in the world. In addition to providing a home and feeding ground for a diverse marine population—including sea urchins, anemones, starfish, crabs, snails, and fish—the kelp has been harvested by man for the extract algin, which is used in medicines, ice cream, and cosmetics, among other things.

Opposite: a sea flea is hoisted aboard.

coast of Villefranche; and a geological program between Corsica and the French Riviera directed by Professor Bourcart.

From September 1 to October 6, Professor Lacombe headed a mission to measure the currents off the coast of Cape Spartel, west of Tangier. On the way back to Marseille, Captain Alinat, in collaboration with Gaz de France, laid several hundred yards of experimental pipes similar to those used for the natural gas project along the route of the proposed Mostaganem-Cartagena pipeline to test the corrosive effects of the sea during long-term use.

Never is *Calypso* occupied with one task alone; rather, many different adventures and studies take place. For example, while concerned with the various missions of 1960 mentioned above, *Calypso*'s team was simultaneously conducting some forty-odd experiments with the diving saucer to determine more accurately the extent of its capabilities and was also testing a newly developed "explosive anchor" that planted itself deep in the ocean floor and expanded to provide more secure anchorage.

To round out the year, *Calypso* sailed back to France. On December 10, 1960, she was in Nice to celebrate the launching of the world's largest inflatable boat, one 66 feet long, 25 feet wide, and equipped with a 480-horsepower engine. One of a class of vessels with extremely shallow draft and great maneuverability, and dubbed Zodiacs by Cousteau, the inflatable craft was designed by him and developed and built by the OFRS. The celebration at Nice included the christening of the Zodiac *Amphitrite* by Princess Grace of Monaco. After the ceremony she and her husband, Prince Rainier, were received aboard *Calypso*.

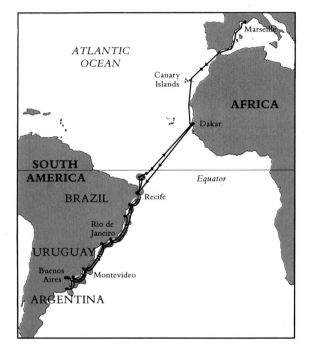

YEAR OF THE SAUCER

The year 1961 proved to be a very busy one for both *Calypso* and the diving saucer she carried in her aft hold. Many diverse scientific organizations availed themselves of the two unique vessels. New milestones were reached during the first month of the year alone: the saucer first reached a depth of almost a thousand feet, at Cape Béar near the Spanish border, in January, and the first woman to descend in the saucer, a biologist from nearby Port-Vendres, was taken on the saucer's fiftieth dive that month.

Calypso spent most of February in the dockyards of Marseille, where she received some much-needed maintenance and repair work. From then on, however, she kept up a frantic pace, occupied by one mission after another in rapid succession. From February 28 to March 5 *Calypso* was involved in many coring programs around Nice, where she was joined by the famous American oceanographic ship the *Atlantic*. From there she proceeded to the area of Port-Vendres and Cape Creus (on Spain's Costa Brava), where, combining the use of the diving saucer and drop-corers, she carried out geological studies. The saucer was used at depths up to 1,000 feet, while the drop-corers were employed at greater depths, some extending to well over 7,000 feet. This mission was followed immediately by a photometry expedition, under the direction of Professor Ivanoff, that took *Calypso* to Sardinia, the coast of Tunisia, and to Sicily, before she returned to Nice via Capri and the Strait of Bonifacio.

Calypso's pace did not let up yet, however. Another geological program with the Edgerton thumper directly ensued, this one in the vicinity of Marseille. Then the ship was off to the clear waters of Corsica to test the efficiency of the diving saucer's horizontal sonar system. Operation Lumen, led by Captain Cousteau, followed straightaway. This was a mission to measure the transparency of open-sea water at various depths. The task was accomplished by lowering a 500-watt lamp and the diving saucer equipped with a recording photomultiplier to the same depth and then gradually increasing the distance between the two until the light from the lamp could no longer be perceived. The results of Operation Lumen indicated that the water was most transparent at a depth of 825 feet, when the light could be seen from as far away as 1,000 feet.

Moored in very deep water of the Atlantic off the Strait of Gibraltar, *Calypso* carries out hydrological work with a companion vessel, the *Hellen Hansen*.

On May 8 *Calypso* sailed for the Strait of Gibraltar to participate, under the direction of Professor Lacombe, in an international program of current measurements on the Atlantic side of the Strait. *Calypso* was joined there by five other oceanographic vessels, but bad weather interrupted and marked the expedition as a rare unsuccessful one for *Calypso*'s team.

Upon her return to Nice, *Calypso* took part in two more missions. The first was a study, along with the *Winaretta Singer*, a ship from the Oceanographic Institute of Monaco, of the marine life populations at the bottom of the area's waters, using the troika to supplement classic oceanographic instruments. Then *Calypso*'s team combined forces with atomic scientists from Saclay, near Paris, to measure the amount of radiation in the water and in sediments between Nice and Corsica. This routine work and the long line of short missions finally came to an end on July 18, when *Calypso*'s

Engine room and mechanical repairs, especially emergency ones, make for hot, dirty work. Here, *Calypso*'s clutch receives attention.

team began making preparations for a six-month-long expedition to South America.

Calypso set off for South America on October 18, in very rough seas, on a voyage beset with mechanical problems. Soon after the ship had gotten under way, the crew noticed that her bow hold was filling with water. After several inspections, it was determined that the water was caused by a leaking pipe. While this in itself seemed a minor setback, the damage it caused was extensive, and since the pipe could not be repaired at sea, the crew had to pump out

Welding is one of the innumerable skills necessary for *Calypso*'s upkeep.

the water continually until *Calypso* reached a dockyard. They set sail for Santa Cruz de Tenerife in the Canary Islands, off the coast of northwestern Africa, arriving there on October 25 only to learn that Santa Cruz had no shipyard. They went on, pumping all the way, to Dakar, where *Calypso* pulled in on October 30.

Once at Dakar, the pipe was quickly and easily repaired, but a more serious problem was uncovered: water was leaking onto the starboard engine. The hole through which the water was flowing was quickly patched with cement, affording a precarious solution at best, and it was under these perilous conditions that *Calypso* began her long transatlantic journey from Dakar to Recife, Brazil—a journey of some eighteen hundred miles.

On November 8 *Calypso* crossed the equator into the Southern Hemisphere. It was her fourth time across and those crew members who had never before crossed the line were ritually dunked in the water by the more experienced members of the crew. Three days later, *Calypso* reached the port of Recife, where she picked up a team of five biologists headed by Professor Jacques Forest. These scientists, Brazilian as well as French, were to aid in the expedition about to be started.

For the next month and a half, *Calypso*'s team took samples, by dredging, trawling, and diving, of various flora and fauna (plankton in particular) in the region between Recife and Mar del Plata in Argentina. Samples were collected from more than 180 different locations, and the ship's many ports of call included Salvador, Pôrto Seguro, Rio de Janeiro, and Montevideo, where she spent Christmas in the company of Jacques and Simone Cousteau, who had just flown in from Europe for the occasion. The multitudinous samples enabled *Calypso*'s team of biologists to discover and identify many new species of marine creatures and contributed greatly to the scientific community's knowledge and appreciation of the area. In fact, before this expedition very little was known about the biology of the Brazilian continental shelf, and virtually nothing at all in the way of data on the ecology of marine communities was available at the time.

After the new year began, *Calypso* embarked on another mission, this time under the direction of Professor Pérès, again taking samples of the various fishes, crustaceans, mollusks, and other fauna and flora that inhabited the region. All these studies took place despite continuing mechanical and weather problems that plagued *Calypso*. Serious mechanical problems began on January 10, 1962, when the main starboard engine failed; then, a few days later, the clutch for the port engine began to show signs of wear; next, after a hard day's work in very rough seas, the crew had to contend with the flooding of the electrical board, which caused a total power shutdown and a great risk of fire aboard ship; next, a rubber hose burst, flooding the engine, and over two hundred gallons of water-polluted oil had to be disposed of. A few days later, the portside generator burned out completely; the same day, an oil leak in the main starboard engine spilled more than fifty gallons of oil into the hold. And so it went. Fortunately, the mechanics and other crew members were able to make the necessary repairs as each bit of bad luck occurred, and the scientists could continue with their work with relatively few interruptions. The mission in the South Atlantic was successfully completed on February 17.

On February 21, after a few of the most major repairs had been made and provisions had been replenished, *Calypso* was deemed ready to tackle the eighteen-hundred-mile Atlantic crossing. She accomplished the crossing safely, until bad luck returned as she approached France. First, storms forced her to seek shelter in a cove off the coast of Spain, and then, as she was being readied to start for home again, she lost her port anchor. Following a long and difficult search, the crew retrieved the anchor, set out again for France, and finally finished the eventful 16,350-mile expedition at Marseille on March 16, 1962.

One would think that *Calypso* and her crew should have been entitled to a long period of rest and repair after so arduous a journey, but such was not the case. Instead they arrived home to find additional programs already waiting for them. Since *Calypso* was the sole official French oceanographic vessel in existence at the time, completion of these missions was left entirely to her. Careening and major repairs would have to wait until November, for no less than sixteen new oceanographic programs would occupy her time from the beginning of May until the end of October.

These new programs took the ship throughout the Mediterranean. They included: sediment coring off the coasts of Monaco and Villefranche; seismic refraction between Nice and Corsica; soundings and troika photography along the French coast; geophysical

studies and inspection of undersea canyons off the coast of the Estérel; seismic profiling off La Ciotat; and hydrology and geology programs all over the Mediterranean.

During the seismic refraction program *Calypso* was often a veritable powder keg, with enough TNT aboard to sink herself and the two ships with which she was working. Explosives detonated underwater created sonic shocks that helped draw a picture both of sea-floor features and the consistency and composition of subsea-floor strata. They had long been used in geology explorations for oil. But this was the first time Cousteau had witnessed the massive use of explosives in undersea geophysics; once he saw the damage they caused the marine life of the area, he began his crusade to have TNT banned from the seas and replaced by less destructive pressure wave generators.

The sediment coring program at Monaco was also particularly noteworthy, as it was carried out with a new radio navigation system, French initials RANA, which enabled the crew to determine *Calypso's* exact position within a few yards, as long as she was inside a radius of two hundred miles from Monaco.

From August 13 to August 29 *Calypso* was on a special acoustics program directed by Professor René-Guy Busnel, head of the Physiological Acoustics Laboratory of the Centre National de la Recherche Scientifique (CNRS) in Jouy-en-Josas outside Paris. The various sounds—clicks, pops, and whistles—made by marine mammals such as dolphins and pilot whales were recorded in the Strait of Gibraltar, where such creatures abound, swimming there on their way into and out of the Mediterranean. Whenever the scientists played back the sounds to another group of mammals they witnessed an amazing reaction: the mammals would break out in an instantaneous stampede!

The next month saw *Calypso's* team involved in one of their most ambitious programs to date: the construction, testing, and inhabiting of the world's first "undersea house," the brainchild of Dr. George Bond of the United States Navy. Bond had conceived of a "saturation method" of diving whereby, to increase the efficiency of the dives and to minimize the chance of decompression accidents, divers would actually live for a period of time below the surface of the sea in gas-filled shelters, or "underwater houses."

Decompression accidents occur when gas from a diver's air tanks dissolves in his blood and later comes out in the form of bubbles in the bloodstream. This results in decompression sickness, which will vary in seriousness depending upon the depth of the dive and the amount of time spent underwater. To avoid problems, each time a diver surfaces he stops at various depths to allow for gradual decompression. This is tedious and costly. The undersea houses of George Bond's invention would allow divers to avoid this decom-

Opposite: Conshelf I's simple barrel construction is tested for leaks before being submerged for the habitat experiment.

Captain Cousteau visits Claude Wesly and Albert Falco during their week-long stay in the comfortable but cluttered quarters of the Conshelf I habitat.

AT HOME IN THE SEA

Conshelf I

pression after every dive. Under the conditions he specified, the divers' bodies would become saturated with gas during their first few hours underwater and remain so for the entire extent of their stay, which could be as long as several weeks. It would then be necessary for them to decompress only once—at the end of their visit. "Pay as you leave," the divers called it.

Dr. Bond recommended his plan to the American navy, but it was rejected, so he turned to Captain Cousteau and his diving team. Cousteau, after due consideration, decided to devote *Calypso* and her divers to the "Conshelf" program, as it was dubbed, and the testing of "Conshelf I" took up most of September 1962. The object of this program, as stated by Cousteau himself, was simple: "Two men living for one week in a small building at a depth of 37 feet and working several hours a day at 65 feet." Conshelf I was a simple, heavily ballasted, room-sized cylinder equipped with an open hatch at the lower end. The cylinder was christened *Diogenes* in honor of the Greek philosopher who did his best thinking submerged in his bathtub.

Conshelf II

74

Calypso towed *Diogenes* to a small bay at the island of Frioul, off Marseille, where, on September 14, the cylinder was carefully submerged at the selected site. Albert Falco and Claude Wesly, the first "oceanauts" in history, swam down to their new home while *Calypso*'s crew, accompanied by many journalists and photographers, followed them on closed-circuit television.

Everything about the Conshelf program went according to plan. The oceanauts left their quarters for several hours each day to work at sixty-five foot depths, then returned to their rather cramped but comfortable new home. To help relieve possible boredom, the two men had the use of a radio and television installed in the cylinder; they also received many visitors, including Captain Cousteau, professor of physiology Jacques Chouteau, and even a newspaper reporter. Every day the divers were subjected to a battery of physiological tests. Falco and Wesly finally surfaced, on September 21, in perfect health, after breathing oxygen for an entire hour to avoid potential decompression accidents. The mission was an unqualified success, Dr. Bond was vindicated, and a new frontier in the conquest of the sea had once again been crossed by *Calypso*.

The rest of the year was taken up by more conventional programs headed by various specialists. These were conducted in the general area between Corsica and Provence. For one week *Calypso* participated in seismic profiles and tests of Professor Edgerton's new "mud penetrator," a device that picked out hard objects buried in sediment. Then the team was off to test the radio chemistry of seawater in various parts of the Mediterranean; from there they tackled a familiar task: hydrology and current measurements. *Calypso* finally returned to Toulon at the end of October. The long-awaited careening and general overhaul could at last be performed, and the ship received a long-overdue four-month rest.

The Conshelf I program had proved so successful that immediately after it had ended Captain Cousteau began hatching plans for an even more ambitious project, Conshelf II, also called "Precontinent II." This second program would demonstrate to the world that important scientific and industrial operations could be undertaken far away from a home base. Conshelf II had three technical goals. The first was to study the effects of an extended underwater stay—an entire month under the sea at 33 feet with work done every day at 60 feet; the second goal was to study the effects of a week-long stay at 82 feet, in a dwelling filled with an oxygen-helium mixture, actual work being performed daily at depths of no less than 160 feet; and Conshelf II's third goal was to install an undersea "garage" for the diving saucer, in which it could be overhauled and reloaded without having to surface. Together the three "buildings" were to make up a sort of underwater village in the Red Sea.

As always, financing the expedition was a problem for Cousteau. No conventional funds could be obtained for so daring and

Swimming beneath the garage of the Conshelf II village, two divers approach the starfish house, the larger of the two underwater dwellings.

expensive a project, so Cousteau turned to his one proven moneymaker, underwater films. He succeeded in obtaining a profitable contract to shoot a feature-length movie, called *World Without Sun*, of the expedition.

Now all that remained to be done before the expedition could actually get under way was the selection of the precise site in the Red Sea on which to set up Conshelf II. On February 27, 1963, *Calypso* sailed for Port Said and the Red Sea with a dual objective: finding the optimum location and shooting for the film some spectacular footage with the diving saucer. The team examined and filmed innumerable reefs in the Red Sea in search of a suitable site. The perfect reef would have tiers that rose from the depths like a staircase, with ledges at 33 and 82 feet. Further, it must be sheltered from the wind and its waters needed to be clear and crowded with brilliant-colored fish suitable for filming. All these conditions were not easy to find simultaneously, but in time the ideal site was located: Shab Rumi, twenty-seven miles north of Port Sudan, the so-called "African capital of coral waters."

Meanwhile, the houses, hangars, instruments, and accessories necessary for Conshelf II were being readied back in France. Once the site was chosen, all these objects were loaded onto an Italian freighter, the *Rosaldo*, specially chartered for the occasion, which set sail for Port Sudan on April 10. *Calypso*'s team used the available time to shoot many striking sequences for *World Without Sun*, visiting dozens of spectacular coral reefs south of Port Sudan in the Strait of Bab al Mandab, the Gulf of Aden, and the Sea of Oman. She

A diver peers into a window of the starfish house.

Below: two aquanauts relax over a game of chess in relatively spacious living quarters.

then returned to the harbor of Port Sudan, where she met up with the *Rosaldo* on April 27.

The items carried by the Italian freighter were mainly prefabricated components which were to be put together and submerged at Shab Rumi to construct the undersea "village." It was the *Calypso* team who were responsible for the actual construction of the houses and the garage. To this end, it was necessary to transport the equipment from one ship to another and take it to Port Sudan, where the pieces would be assembled. The *Rosaldo* set up a permanent anchorage inside the lagoon at Shab Rumi, and from there *Calypso* made fifteen trips back and forth to move the equipment, prepare her own anchorages, and tow out to the site the assembled cylindrical houses and bright yellow garage. Submerging the structures themselves was an extremely arduous task. More than a hundred tons of lead cast into one-hundred pound weights had to be positioned by the divers.

On June 12 the oceanauts who were to live in the larger "starfish house" (so named because workshops and living quarters branched off from the main chamber) at 33 feet—Albert Falco, Claude Wesly, Pierre Guilbert, Raymond Vaissière, and Professor Pierre Vanoni—took possession of what would become their new home for the next month. Canoë Kientzy and André Portelatine moved into their "little house" 83 feet below the surface on July 5.

During the span of this experiment, the oceanauts worked under strict medical control, monitored through a control room set up on board the *Rosaldo*. Although they carried out daily biological experiments in their undersea laboratory and worked every day at depths (for the inhabitants of the large house) of 69 feet and (for the inhabitants of the smaller house) of 160 feet, they also had time for

Components of the saucer garage wait to be assembled.

more relaxing pursuits. They were exposed to sunlamps daily, played chess, listened to music. They had a noisy parrot for a pet, and Guilbert the cook even trained a large triggerfish. They also received many visitors, most notably Dr. Charles Aquado of the United States Navy and Philippe Cousteau, then twenty-two years old. The only side effects felt by the oceanauts were by those in the deeper house, who safely made dives much deeper than had been anticipated—sometimes reaching as far as 330 feet—but suffered on those occasions from excess moisture. The men of the Conshelf station did notice several subtle effects of the pressure at such great depths, however: cuts and abrasions healed faster than when above the surface, for example, whereas beard growth was noticeably slower. Interestingly, Professor Vanoni, who on land suffered from claustrophobia, did not feel uncomfortable at any time in his undersea house.

The starfish house—the main unit of the Conshelf II complex—floats alongside a dock at Port Sudan while the diving-saucer garage is prepared for submersion.

The oceanauts returned safely to the surface on July 14, Bastille Day, to a great and triumphant welcome. But soon the dismantling operations had to begin, and once again *Calypso* shuttled back and forth between Shab Rumi and Port Sudan. The saucer garage was not dismantled, but left behind for possible reuse.

Together with the *Rosaldo*, *Calypso* left Port Sudan on July 25 and sailed to Marseille, where she was quite warmly received. *Calypso*'s team had proven the feasibility and usefulness of extended underwater stays. In addition, the film *World Without Sun*, a by-product of the expedition, won for Cousteau both international acclaim and his second Academy Award.

After a ten-day rest, *Calypso* resumed her scientific activities. These seemed almost routine after the great adventure in the Red Sea. At the end of August she participated in the examination of a site proposed for the disposal of waste from an aluminum factory at Gardanne through a three-mile-long undersea pipeline. Then, in September, the crew sailed to the island of Alborán to film once again the kelp forest there that had so intrigued them. A hydrology program in the Nice–Corsica area took up the month of October, and was followed by a biology program to study the deep benthos populations between Cassis and St. Tropez. *Calypso*'s year ended with more saucer dives in the vicinity of Port-Vendres and a geological program and study of radioactivity off the coast of Nice. *Calypso* finally returned to her home port at the end of December, after what had been one of her busiest and most productive years to date.

Calypso's schedule did not let up in 1964. In fact, so many projects had been arranged for her that repairs traditionally made during January had to be limited to those deemed absolutely imperative.

Her first mission consisted of taking part in towing and anchoring a new laboratory-buoy, nicknamed the "Mysterious Island." This odd-looking structure was conceived by Captain Alinat and developed and produced by the OFRS under the supervision of Jacques Picard. It consisted mainly of a 200-foot-long anchored tube that would permit for the first time permanent and continuous research at sea. The tube, which was to float vertically in the water, was designed so that research could be conducted there in a stable and efficient manner under even the most adverse weather and sea conditions.

The need for such a perfectly stable platform on which to work was felt simultaneously by the *Calypso* team and by the Scripps Institute of Oceanography in La Jolla, California. The latter developed a similar, but larger structure nicknamed "Flip" because it had to be towed horizontally and then "flipped" into its vertical position at the site of the operation. The major difference between the two laboratories was that Flip was a drifting buoy and the Mysterious Island was to be anchored.

Calypso made her way to the harbor at Villefranche-sur-Mer, where she met up with the Mysterious Island and a French navy ship, the *Marcel Le Bihan*, which was to tow the laboratory-buoy to the anchoring spot, halfway between Nice and Corsica. On January 21, 1964, the towing began, at such a slow speed that it took four days to reach the intended destination. Nevertheless, the Mysterious Island was delivered safely and uneventfully, and the heavy moorings, as well as the two-mile-long nylon-and-Dacron-braided anchor line, were installed without difficulty. The floating island was then ready to be turned over to the scientists who were to work aboard it for ten years, exploring firsthand the mysteries of the sea.

Once the Mysterious Island had been transferred into their hands, *Calypso* was able to resume less spectacular but equally important scientific programs, such as making drillings and corings off Cap-d'Ail near Nice, hydrology programs, current and wind measurements, and inspection of the waste dumpings at Cassis. This routine work was interrupted on March 17, when the ship was called back to Toulon to undergo a major overhaul.

MYSTERIOUS ISLAND

The ship and her crew were back at sea in a month. Saucer dives were resumed on April 18, when Canoë Kientzy became the first man other than Albert Falco to pilot the SP-350. Kientzy took scientists from Dr. Pérès's team down to inspect the area off the coast of Crete in a biology program that lasted for some six weeks.

On June 2 *Calypso* began a seismic-reflection program in the western Mediterranean. The testing there of new equipment—sparkers, boomers, and air guns—was interrupted on June 26 so that *Calypso* could take André Maréchal, then head of the French national scientific research program, to inspect the facilities of the laboratory-buoy floating in the bay of Villefranche-sur-Mer. This became in itself quite a media event, as reporters and cameramen from newspapers and television crews were on hand to document the visit.

Calypso sailed on June 29 for the Italian port of La Spezia, where she took delivery of an instrument that would become indispensable to deep-water diving. The Galeazzi SDC (for submersible decompression chamber) made it possible for a diver to remain dur-

Before being towed to its site, the laboratory-buoy is tested in the shallow bay of Villefranche-sur-Mer.

ing his ascent at the same pressure as the depth at which he had been working. Then, once back on board, inside the chamber he could be decompressed gradually and under medical supervision. The instrument was found to do double duty: while carried aboard *Calypso* it served as an observation tower from which the crew could navigate and also search out some of the larger forms of marine life. It came to be called simply the "Galeazzi tower."

From La Spezia, *Calypso* sailed to Messina to participate in a detailed geophysical and geological study in the Strait of Messina. The object of the study was to determine the feasibility of the Italian project to build a high bridge connecting Sicily with mainland Italy. Once this mission was completed, *Calypso* became involved in another program in Sicily: an optical study of the area around the Stromboli volcano in the Lipari Islands northwest of Messina. August, September, and October were taken up by more routine expeditions, saucer dives, and seismic-refraction programs throughout the Mediterranean, and the period ended with still more studies in the Strait of Messina in connection with the Sicilian bridge project.

From October onward, *Calypso*'s team were devoted to perfecting deep-diving methods in preparation for the operation that would be attempted the following year: Conshelf III. They initiated a series of training dives ranging from 260 to 330 feet, using the Galeazzi tower and a helium-oxygen mixture. Accustomed to suffering from the usual dizzying symptoms of the so-called "rapture of the depths," the divers were delighted to find out they could remain clear-headed when breathing this "heliox" mixture. Their relief was tinged with disappointment, however, for, as Albert Falco remarked, "On air, we find everything so beautiful, but with heliox, the reality is there, gray and sad." Without the heliox, however, surfacing had been a long and tedious operation.

Two teams of six oceanauts each were trained in these deep-diving programs off the coast of Villefranche-sur-Mer. Tests here lasted for about two months, and by December 15 hundreds of dives had been made, proving the desirability of the Galeazzi tower and perfecting the heliox mixture. *Calypso* could finally return to Marseille for her winter round of repairs.

The year 1965 started out, for *Calypso*, with tragedy. At seven o'clock on the night of February 20 she received an S.O.S. with the message that the Mysterious Island had caught on fire. The entire team gathered and mobilized within hours and were joined in Marseille by Captain Cousteau, who flew in from Paris that same evening. *Calypso* then headed for Nice at top speed to aid in the rescue operations.

By the time *Calypso* reached the scene, however, the fire had already caused extensive damage to the buoy. It had only taken a few hours to reduce the world's first floating marine laboratory to an empty, burned-out shell. The Mysterious Island, which had successfully carried out twenty-one missions within thirteen months,

Captain Cousteau looks back at the laboratory-buoy after fire destroyed its interior.

was now seemingly destroyed forever. Luckily, the buoy's six occupants—François Varlet, Pierre Oriol, J. P. Rebert, J. P. Bassaget, Claude Wesly, and skipper Gabriel Mariani—had all been safely evacuated. They had jumped 60 feet down into the black water in the middle of the night to reach a small inflatable dinghy, from which they were rescued by the *Alizée*, a small freighter that had been in the area.

The loss of the Mysterious Island would have been a great one indeed. But after inspecting the laboratory, Cousteau felt that it was possible to salvage it. *Calypso*'s team sailed for Marseille confident that the laboratory-buoy would be reconstructed. After several months work, it was put back into operation and remained to serve the oceanographic scientific community for eight more years.

Once the drama of fire on the high seas was over, *Calypso* returned immediately to her more usual duties. The year's first expedition took her to the mouth of the Rhône and off the coast of Nice for a hydrology program led by Professor Lacombe. Then the ship and her team were off again, this time on a geophysical and acoustical mission with Professor Muraour along the southern coast of France. From there they traveled to Tunisia and Libya, where, from April 13 to May 10, they participated in a biological and geological survey. This expedition, directed by Jacques Picard, involved photographic profiles with the troika, hydrological studies, corings, soundings, dredgings, plankton hauls, and saucer dives. Then followed an undersea optical program that took them to many different areas in the Mediterranean.

Calypso devoted five periods of two weeks each during July, August, October, November, and December to making an accurate mapping of the zone covered by the RANA radio navigation system—from Italy to Toulon, including most of Corsica. The results of these investigations enabled the most detailed topographical maps and charts of the area to be printed.

The periods from July to December that were not occupied in mapmaking were nevertheless not times of rest for *Calypso* and her crew. Rather, these witnessed many other investigations, including studies requested by the CNRS. These included a study of deep currents in the Cadiz–Casablanca area; seawater radioactivity measurements in the western Mediterranean; seismic profiles off the coast of Monaco; and saucer dives near Nice and Marseille.

The grandest project undertaken by the *Calypso* team in 1965, however, was operation Conshelf III, also called Precontinent III. Conshelf III was put into play 328 feet below the surface of Cape Ferrat off the coast of Villefranche; at the same time a similar experiment, the United States Navy's Sealab 2 project, was being conducted some 200 feet down, off San Diego in California.

Conshelf III was a spherical structure, 20 feet in diameter, equipped to accommodate six men safely and comfortably for the twenty-seven days of the experiment. These men—Christian Bownia, André Laban, Philippe Cousteau, Raymond Coll, Yves Omer,

and Jacques Rollet—entered the sphere in the harbor of Monte Carlo, where they were submitted to eleven stages of helium-oxygen mixtures, simulating the pressure of progressively greater depths. The sphere was then towed by *Calypso* to Cape Ferrat and slowly submerged at the chosen site, where it was connected by power and communication cables to the Cape Ferrat lighthouse. While underwater, Rollet compiled medical, physical, and physiological data on all six oceanauts. This information was then fed into a computer on shore. The oceanauts swam daily down to depths of 394 feet to perform work on a mock-up oil wellhead that had been installed there by *Calypso*'s crew.

One evening, a telephonic connection was linked up between the oceanauts of Conshelf III beneath the Mediterranean and the aquanauts of Sealab 2 deep under the Pacific. In the characteristic Donald Duck–like voices caused by breathing helium, Philippe Cousteau and aquanaut-astronaut Scott Carpenter exchanged greetings and good wishes.

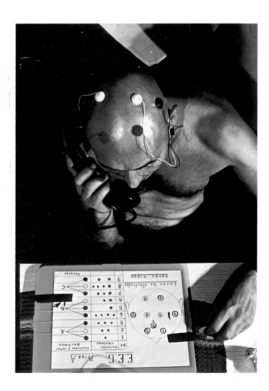

At the end of the Conshelf III experiment, the six deep divers were towed back in their still-pressurized sphere to Monaco, where they were gradually and safely decompressed over a three-day period.

Conshelf III had been both extremely successful and valuable, but the cost of the operation was enormous and caused *Calypso*'s

Opposite, above: safely returned to the habitat, three of Conshelf III's six deep divers check their equipment after a daily work stint at depths of nearly 400 feet.

Opposite, below: wearing a garland of EEG electrodes, André Laban performs one of the battery of physiological tests given daily to the Conshelf III aquanauts.

Below: aquanauts prepare to enter the spherical chamber that will be their underwater home for the next four weeks of the Conshelf III program.

team to go into great debt. A one-hour television special, made largely with films taken by Philippe Cousteau during the expedition, helped to defray some of the costs. This special, which was filmed for American television under the auspices of the National Geographic Society, would have a profound influence on *Calypso*'s destiny in the years to come.

All in all, 1965 was a very full year indeed for *Calypso* and her teams of divers, scientists, and crew. On December 18 the ship returned to Marseille, where she was met by Vincent Bianco, who would perform the extensive repair work necessary.

Freshly painted and newly renovated, *Calypso* sails by the elegant headquarters building of Monaco's Musée Océanographique.

CALYPSO GETS A FACE LIFT

The year 1966 marked a new orientation in *Calypso*'s activities. For the previous fifteen years she had been the only French oceanographic vessel, always on call somewhere. Scientists, researchers, and private institutions constantly clamored and competed for her services. That was changed in 1966, however, by the introduction of a brand-new French oceanographic ship, the *Jean Charcot*. *Calypso* was thus freed from some of her more routine duties to concentrate on more adventurous and esoteric ones. Perhaps equally important to her evolution, the film shot during the Conshelf III operation had proved to be a huge success on American television. It led to a contract for a series comprising an additional dozen films, and this provided for *Calypso* and her crew almost complete financial autonomy during the next three years.

This lessening of the work load, combined with the increase in available funds, allowed *Calypso* to undergo a major program of refitting and remodeling. The overhauling work was carried out in two separate stages, the first in January and the second throughout October, November, and December.

When the first stage of repairs had been completed, *Calypso* began a series of geological studies of structures that occurred at great depths and appeared to be related to salt mounds. These structures were first discovered by the *Calypso* team in 1957 and had intrigued them ever since. To find out more, they now set up some technologically improved seismic-reflection equipment in the Nice–Corsica region, an operation that took them until April 20 to complete. This was followed by an even more extensive seismic operation, one in which *Calypso* was joined by the *Jean Charcot*. Between them, these two oceanographic vessels covered the entire western Mediterranean basin—from Nice to Corsica, Sardinia, Majorca, and the Spanish and Algerian coasts. Also, at Cassis, *Calypso* supervised the laying of the controversial underwater pipeline that would deposit in deep water the famous red-mud waste from the aluminum factory in Cassidaigne.

No sooner had these projects been completed than representatives from a foreign airline asked the *Calypso* team to quickly locate and bring to the surface the wreckage of a plane that had just crashed and sunk somewhere off the coast of Nice. They were able to locate the wreck the first day of the search, but as the diving saucer descended it triggered an avalanche of mud, partially burying the plane. It took two weeks of very hard work and difficult maneuvering to excavate and examine the wreckage, but several important pieces were recovered, pieces that enabled the airline's representatives to reconstruct the causes and conditions of the accident.

For the rest of her active year, *Calypso* was devoted to many separate scientific programs, all in the general area around Toulon. From July 4 to August 4 she was occupied in testing several new instruments that recorded the optical properties of seawater. The month of August was largely taken up by saucer dives and use of the troika in geological programs. September saw still more saucer dives, most notably on September 12, when the mayor of Marseille, Gaston Deferre, went down in the saucer to inspect the site of the controversial red mud. Finally, from October 1 to 11, *Calypso* finished her year by participating in a seismic-reflection program in which the troika monitored the results.

At last, in mid-October, *Calypso* was freed from her duties and enabled to return to Marseille for the second stage of her long-awaited renovation. Immediately upon arrival, the ship was careened and stripped. Her barges were sandblasted; the smokestack was removed, cleaned out, and remodeled; furniture was repaired; and engines, generators, mufflers, and rudders were all sent out to shipyards for overhauls. The winch was removed and rebuilt with two separate drums. The davits were removed and reconstructed to support a weight of one and a half tons, and a 25-kilowatt rotary converter was added to energize *Calypso*'s new precision instruments. Modern transformers, able to hook up to shore lines when *Calypso* was in harbor, at last were installed to replace the old-fashioned main electric battery. The old "false nose" was replaced by a newer model with a larger observation chamber and eight portholes for better visibility. Even the deckhouse was entirely re-modeled; the crew's quarters were rebuilt into separate cabins to accommodate two persons each.

The ship was floated on December 28. All the renovation and remodeling had been completed, and the only thing left to do before *Calypso* could get on her way again was to equip the quarterdeck for new equipment soon to be brought aboard: a Galeazzi tower for deep diving and two new one-man P-500 minisubs, nicknamed "Sea Fleas." These Fleas had been designed by the OFRS according to Captain Cousteau's specifications. They were small craft, no more than three feet wide and six and a half feet long, and weighed about two tons. Smaller than the SP-350, they were also safer, for these bright yellow saucers were built so that they could latch on to one another for enough power to tow a debilitated saucer if the case should arise. Each had two water-jet engines in the forward section and an electric motor enclosed in a fiberglass compartment attached to the outside of the hull.

A sea flea, *Calypso*'s one-man minisub, is observed through a fish-eye lens.

THE GREAT EXPEDITION

On February 18, 1967, Jacques Cousteau held aboard *Calypso* a press conference in which he unfolded to journalists the program that would occupy his team for the next four years. This four-year period would be devoted largely to the filming of the twelve-part series called *The Undersea World of Jacques Cousteau* for American television, the subjects of which had been left entirely to Captain Cousteau's discretion. During the press conference, Cousteau emphasized the destruction already witnessed in the world's oceans caused by overfishing and pollution, and stated his intention to record on color film for future generations the vanishing marvels of yet-unspoiled waters. After the conference, a farewell reception, attended by Princess Grace and Prince Rainier of Monaco, among others, was held aboard ship.

Later that day, after much celebration, during which many hands had been shaken and hundreds of balloons had been released, *Calypso* sailed from Monaco for the Red Sea, the first leg of her longest and most ambitious expedition. Whereas up to this point *Calypso*'s voyages had been specific and definite in scope, this time she was off on a trip destined to be more spontaneous, more adventurous, and less structured than any that had preceded it. This one was viewed as an expedition without limits of time, space, or purpose—a sort of permanent mission, the object of which was simply to explore the secrets of the sea, no matter where those secrets might be found or what they might reveal. For once, *Calypso*'s mission, like the sea itself, seemed boundless and unending.

This is not to suggest that Cousteau and his team had no operational plan; on the contrary, their plan was the most comprehensive ever prepared, and prepared with the scientific support of the Musée Océanographique de Monaco, of which Cousteau had been a director since its founding in 1957. The plan was essentially to become real residents of the sea, to seek knowledge wherever it might be found, and to film the many beauties and curiosities of the under-

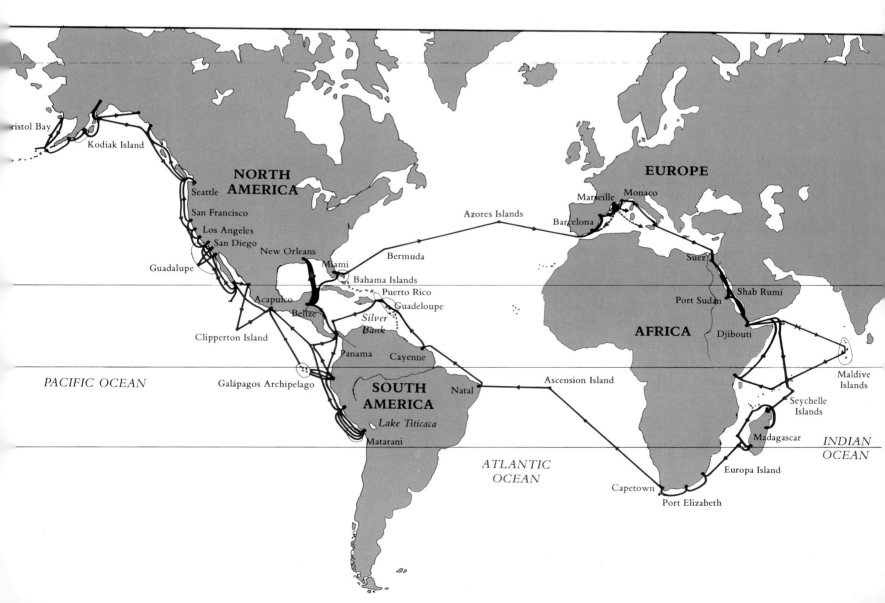

sea world. The results garnered would be completely true-to-life, for the expedition would be documented by researchers and scientists from the Musée Océanographique. During the making of each film *Calypso* would carry an undisputed international authority to supervise the accuracy of the work and the team's observations.

Also on board, in addition to the scientists from the Museum, were Frédéric Dumas, Cousteau's old diving partner, and the first of two diving teams. Two separate diving teams were necessary for this expedition because maritime law stipulated that divers could only work for six months at a stretch. Captain Cousteau was aboard, of course, despite being in pain from two broken vertebrae suffered in a recent automobile accident; as always, Simone Cousteau was ready to sail. The newest passenger aboard *Calypso* was Zoom, a Saint Hubert hunting dog presented to Simone by Princess Grace at the farewell reception in Monaco. Zoom soon became a favorite on board *Calypso*, although his large size and infinite energy seemed at first a bit excessive for the ship, which was rather cramped for space due to the considerable amount of equipment taken on for the voyage.

Largest among these pieces of equipment were the Galeazzi chamber and the SP-350 two-man diving saucer, both heavy and cumbersome, and requiring for their operation another large piece of machinery, the crane Yumbo. Of course, navigational aids such as radar, sonar, and automatic pilot systems also took up space. In addition, there were numerous small craft aboard, all of which were needed for different types of reconnaissance, observation, and protective work. Thus there were runabouts, Zodiacs (inflatable rafts), and inflatable skiffs, plus a large collection of outboard motors. Also, of course, there were all the pieces needed for undersea and above-surface filming: fifteen underwater cameras conceived, de-

A single-person propulsion vehicle, or "scooter," greatly facilitates divers' underwater work.

Opposite: in an experiment to test the reactions of sharks to divers fully dressed in wet suits and with aqualungs, the *Calypso* team created a dummy—nicknamed Arthur—by stuffing a metal-framed suit with foam rubber and pieces of fresh fish.

veloped, and built by Cousteau's specialists at OFRS; six commercial land cameras; and an untold amount of waterproof flashes and electrical wiring. Finally, there was the diving equipment itself. Included in this were newly designed, self-contained, streamlined diving suits with headpieces and aqualungs attached. The headpieces consisted of bright yellow helmets with built-in radio and sonar telephones, searchlights, emergency buzzers, and even movie cameras. The new streamlined aqualungs were of the same bright yellow color, incorporated antishark clubs, and were designed to increase the divers' speed. The suits themselves were black with yellow stripes, which aided in visibility as well as looking very elegant and stylish. The divers also had new motorized scooters which greatly increased their mobility in the sea.

During the past fifteen years the *Calypso* team had acquired a vast oceanographic background through its close association with the top scientists in various fields. Now came the time to put all this experience into actual practice. The aim was to inform a wide public of the dangers that loomed ahead, threatening the very survival of life below the surface of the seas.

Soon after her departure from Monaco, *Calypso* entered the Red Sea, where she met up with the *Espadon*, a small tender that had paved the way for *Calypso* for several months. Philippe Cousteau and Albert Falco had been aboard the *Espadon*, helping to locate the sites in the Red Sea most favorable to diving and undersea filming.

After many dives in the Red Sea, *Calypso*'s team headed for the Indian Ocean, reaching the island of Socotra off the horn of eastern Africa on the thirteenth of March and sailing onward to the Maldive Islands, southwest of Ceylon. On their journey the team came across a pod of sperm whales, or cachalots, and marked the spot for their return.

Along the way to the Maldives, on March 18 at two o'clock in the morning, a loud cracking sound filled the air and one of *Calypso*'s twin engines began to race. The shaft of the rear propeller had just snapped, and the ship was brought to an immediate standstill some eighteen hundred miles from the nearest major port. Albert Falco dove below to check the damage, and then, with the help of Christian Bonnici and Raymond Coll—in the middle of the night—managed to secure the shaft and propeller so that they would not harm the rudder. *Calypso* was able to resume her journey, although she limped on toward the Maldives under the power of a single engine and at an exceptionally slow speed.

On March 21, *Calypso* arrived at Lari Atoll, one of the Maldives' many atolls. During the exciting weeks that followed, the Cousteau team explored and filmed the hauntingly beautiful coral and other underwater life of these exotic islands, where the inhabitants lived in houses made from blocks of coral, and where the chief—indeed, only—industry was the raising of coconuts.

Many spectacular underwater creatures were discovered and filmed there, just as Cousteau had anticipated in such a lush coral sea. Besides the corals themselves, rising from the ocean's floor, divers found great pink sea fans, to which bronze-colored feather stars had attached themselves; bright yellow, red, violet, blue, and green sponges; many brightly colored tropical species such as parrot fish, surgeonfish, trumpet fish, dragonfish, and butterfly fish, in vivid patterns of stripes, dots, and other shapes; and strange "garden eels," which lived buried in the mud and emerged from it shaped into curved question marks—creatures that Cousteau had first encountered in Madeira in 1948.

A diver, equipped with underwater camera and flash, films a starfish in the Mediterranean Sea.

Opposite: alerted to the presence of an alien being, an Indian Ocean crustacean bristles in defensive posture.

From the relative security of his steel cage, and holding a barracuda as bait, a diver plants a bright orange tag just behind the dorsal fin of a reef shark in order to facilitate study of that species's migratory patterns.

Calypso, still running on only one engine, left the Maldives on April 7 and made her way to a variety of exotic ports in the Indian Ocean. Among these were Mahé in the Seychelles, the Cosmoledo islands—still a wild paradise—north of Madagascar, and Pemba Island, near Zanzibar, to name just a few. Along the way, of course, the ship made many moorings, and her team of divers went below the sea's surface hundreds of times in search of striking marine creatures that would add exciting footage to the television series.

On May 5 *Calypso* arrived at Mombasa, Kenya, where she at last came across a properly equipped shipyard. While there, she underwent extensive repairs, including the installation of a new propeller shaft. This completed, she sailed, once again under the power of both engines and at normal speed, away from Mombasa and back toward the Red Sea.

One day, en route to Djibouti, *Calypso*'s alarm sounded and an unidentifiable large black form appeared portside. Falco, Barsky, Michel Deloire, and Coll set out in a Zodiac to investigate, and discovered that the form was in fact an inoffensive but enormous whale shark, almost 40 feet long. An encounter with such a shark, by far the largest and heaviest fish in existence, is a very rare occurrence, so the cameramen, extremely anxious to capture it on film, slipped unhesitatingly into the water. Deloire succeeded in filming it both in profile and head-on, while Coll, grabbing onto its tail, managed to catch a ride with the beast. Suddenly they were joined in the water by yet another whale shark, this one even larger than the first—maybe 50 feet long—and Deloire turned his camera on this colossal specimen. Again Coll held on, this time to the shark's huge dorsal fin, and dove with it to a depth of almost 150 feet. Coll later reported that at no time did either whale shark react to him; neither tried to escape his grasp or harm him, even when he pierced the sharks' skins with his spear to mark them with plaques that would aid in tracking their migration.

Calypso pulled into Djibouti on May 25, and sailed from there through the Strait of Bab al Mandab, braving both sandstorms and monsoons. Divers and scientists devoted themselves to the study of sharks in the Red Sea for the next few days, filming extensively among the coral reefs of the area. When *Calypso* arrived in Port Sudan, on June 1, the *Espadon* was there to meet her.

On June 3, the entire team paid a visit to Shab Rumi, the site of the Conshelf II undersea village experiment. Divers went down to inspect the site and the saucer garage that had been left behind in 1963. Coral had already begun to grow on the roof of the garage; it would not be long, they concluded, before the entire area would be covered with coral, and the sea would have wiped out every reminder of their presence.

While at Shab Rumi, Cousteau and his team received radio reports stating that relations between Egypt and Israel were growing increasingly strained. War was imminent. Cousteau resolved to return as soon as possible to France, fearing that *Calypso* might be held up by the hostilities. He managed to catch a flight to Paris on June 5, the first day of the Six-Day War.

Meanwhile, *Calypso* and the *Espadon* sailed to Suez, where they remained, anchored side by side, for the duration of the war. The *Espadon* was manned by a skeleton crew of only two men; *Calypso* carried a full complement, and no one was allowed ashore. It was a terrifying time for *Calypso*'s team: they were caught in the middle of crossfire, and bombs dropped all around the ship. No water or food was allowed to be brought on board, but the Egyptian authorities did, after deliberation, agree to permit some of *Calypso*'s precious cargo—underwater scooters, cameras, film—to be unloaded and forwarded to France. Unfortunately, once on the dock the crates with the equipment were hit by machine-gun fire from Israeli planes. The contents were destroyed.

Although the official ending of the war came on June 10, hostilities continued for some time afterward, and the Suez Canal remained closed to all traffic. On July 14 the oil refinery at Suez was bombed, and *Calypso* was once again sitting in the middle of a battlefield. Shell fragments fell all about her. As this day was also Bastille Day, the anniversary of the French Revolution, the crew of *Calypso* had planned a fireworks display to celebrate the occasion, but their festivities were lost in the overpowering sound of gunfire, the smell of burning oil, and the black clouds of smoke that hung overhead.

Hostilities finally ended for good on July 21, and *Calypso*'s crew members were allowed to disembark for the first time since the war had begun on June 5. A completely new team, this one led by Captain Alain Bougaran, arrived to replace the Roger Maritano team, with Canoë Kientzy replacing Albert Falco as head diver. Since the Suez Canal would be closed indefinitely, however, *Calypso* could not yet return directly to France; instead, the crew devised a program in the Indian Ocean that, although improvised, would help make up for the two months of valuable time lost because of the war.

Calypso sailed out to the Indian Ocean, leaving the *Espadon* behind to return to Europe around the Cape of Good Hope, on the back of a German freighter. While sailing to their first planned stop, Djibouti, the *Calypso* team made hundreds of dives, as much to try out the new team members and give them additional practice as anything else, but several miles of film were garnered along the way.

Since this was in large part an impromptu phase of the expedition, the diving team managed to fit into their schedule a visit to a saltwater lake—Lake Assal, in the vicinity of Djibouti—which had intrigued them for some time. Leaving *Calypso* behind, of course, they traveled by helicopter over miles of barren but starkly beautiful land to reach the lake. Once they had arrived, they immediately suited up and waded into the water. Much to their surprise, however, they found the water in the lake so high in salt content that they could not submerge themselves; the amount of ballast normally used in the open sea was not nearly sufficient to weigh them down. After several unsuccessful attempts, they split up all their ballast between just two divers, Philippe Cousteau and Serge Foulon, who, thus equipped with over sixty pounds of lead weights each, managed to descend.

The trouble it took to get beneath the surface proved worthwhile, however, for the lake's bottom was covered with beautiful gypsum crystals that sometimes adopted the shape of flowers and that provided a striking background upon which they filmed sequences for the television series. In addition, the divers found to their astonishment a school of coral fish, probably the only group of coral fish in the world to inhabit inland waters. Unfortunately, they also found that the extreme salinity of the water resulted in a cruel irritation of their skin, which broke out in rashes so severe that it was necessary to hose down the divers with fresh water between dives in order to give them some relief.

Soon after they had completed the program at Lake Assal, the team rejoined *Calypso* off the coast of Djibouti, where on September 22 they began the testing of a hot-air balloon, a new addition to the ship's complement of exploratory devices. Captain Cousteau had been pondering for some time the necessity of an aboveground vantage point to serve as a camera platform and for navigational purposes. Because a helicopter was too expensive for many missions, Cousteau had hit upon a hot-air balloon as the solution. A few months prior to this date Philippe Cousteau had been sent to South Dakota to learn to pilot the machine. The time had now arrived for him to show off the balloon and his flying ability to their best advantage.

Philippe stepped into the gondola of the brightly colored balloon and, as the crew members watched in awe, began to ascend majestically. Without warning, however, the balloon dropped precipitously into the sea, dunking Philippe to the great amusement of the crowd. Just as suddenly, the balloon took off again, picked up altitude, and was eventually brought under control by Philippe. Apparently there were still some bugs to be worked out of this new airborne filming platform; but soon these would be overcome, and the balloon would prove to be of great value in both filming and guiding *Calypso* through intricate coral formations when it seemed as though she was hopelessly trapped.

When the balloon testing was completed, *Calypso*'s team once again turned their attention to further exploration. While sailing back out toward the Indian Ocean they encountered several pods of sperm whales, which they studied and filmed in the Gulf of Oman. Experiments revealed the great intelligence of these huge sea mammals; the contact also confirmed their friendliness and curiosity, which sometimes caused the whales to be too slow to retreat at the first approach of ships and to become injured as a result.

Back at the Seychelles, *Calypso*'s crew spent much time diving among the beautiful coral formations, observing and filming sea turtles and the strikingly colored coral fishes, groupers in particular. Groupers have always held a particular attraction for the *Calypso* divers because these fish never appear frightened by the presence of man; rather, they seem angry that their private sanctums should be invaded. On this expedition, however, the divers managed to establish cordial if not truly friendly relations with the groupers, feeding them by hand and petting some. The men knew, nevertheless, that what appeared to be affection was really hunger; the fish appreciated the food offered more than the men's attentions.

In November *Calypso* docked in Madagascar for careening and maintenance. While she was there the two Sea Fleas were delivered for more complete testing. *Calypso* set sail from Madagascar for the Glorieuses Islands, between Madagascar and the Comoros, where the Sea Fleas were tested extensively in surroundings that were indeed glorious. The Fleas proved a real boon to deep-sea exploration; they were light and easily maneuverable, and introduced the concept of the buddy system into the use of minisubs, for they worked in pairs.

At the end of the year, the Bourgaran team was relieved by the Maritano team, which had been on board at the outset of the expedition.

The beginning of 1968 saw *Calypso* still in the Indian Ocean, unable to return to Europe via the Suez Canal, traveling from island to island. At the Comoro Islands, Cousteau's team carried out an obstinate but unsuccessful search for coelacanths, the famous "fossil fish" which had been thought extinct for some seventy million years until a living specimen was discovered near Madagascar in 1938. From the Comoros, *Calypso* sailed south, first to the Bassas da India Islands, a rare example of an Indian Ocean atoll covered entirely by a few feet of water, then on to the island of Europa, where she pulled in on January 9.

The model of patience and gentleness, a green sea turtle allows a diver to touch and even handle it, apparently indifferent to man's presence.

Above, right: in an experiment designed to test a fish's reaction to an intruder of the same species in its territory, a grouper reacts to its reflection in a mirror held by divers. Thinking he had spotted a rival, he would later charge and break the mirror.

The purpose of *Calypso*'s visit to Europa was twofold. First, the island's beaches were known to be a favorite mating and egg-laying spot for sea turtles. It would be a perfect site for the filming of the amorous adventures of these enormous amphibians, some of which weighed as much as 450 pounds and which came from as far away as the Gulf of Aden. Second, like so many other coral islands, Europa had an intriguing underwater channel, the site of sea level during a recent glacial period. (The level of the sea rises and falls with each glacial period; whenever the level remains stable for a few millennia, a bank or beach results on the shore, fostering considerable shallow-water life.) Europa's channel ranged from 345 to 360 feet below the surface, with cliffs and grottoes above the channel and more "fossil beaches" and caverns at 150 feet. The exploration of Europa's ancient submerged shoreline proved arduous work because for every ten minutes spent at 360-foot depths the divers had to undergo two hours in the decompression chamber. In this instance, as in so many others, the results proved well worth the effort. The grottoes were indeed enchanting; their walls, lit up brilliantly by the cameramen's floodlights, were covered with bright blue sponges, sparkling tunicates, and red and yellow sea fans; the ceiling was literally dripping with lilac-colored sea anemones.

While moored at Europa, *Calypso* rode out several severe storms; this was typhoon season in the Indian Ocean. In fact, the ship was directly in the path of two typhoons, named "Flossie" and "Georgette" respectively. Flossie, the first to hit the area, passed so close to Europa that *Calypso* was forced to move quickly out to sea, stranding a team of cameramen on the island while 160-knot winds raged about them. A message was sent from the ship to the men on the island telling them to bury themselves in the sand and wait out the storm; luckily they followed these instructions, for their camp was destroyed by the fierce winds and rains.

As soon as Flossie passed over, Georgette followed in its wake, to the astonishment of meteorologists who had been tracking the storm. It had already caused the death of twenty-three persons on the African coast before it turned suddenly around and headed straight for Europa Island. The strong winds made it impossible for *Calypso* to return to the island to pick up her team, and to complicate matters further, as the ship sailed to Tuléar on Madagascar's southwestern coast, where she could have found shelter from the storm, her starboard propeller shaft snapped, debilitating her completely. To get the ship moving again, three divers—Christian Bonnici, Raymond Coll, and Bernard Chauvelin—dove in ten-foot waves and gale winds and worked underneath the ship for two hours to repair the shaft. This accomplished, the ship fled to Tuléar to wait out the storm. She later returned to retrieve the men left stranded on the island of Europa.

Their work in the Indian Ocean completed for the time being, the team began in February 1968 to wend their way, via the Cape of Good Hope, across the Atlantic and toward the Caribbean Sea, where a difficult marine archaeology program awaited them. Along the way, as was their wont, they stopped at many different ports and islands to study the marine and oceanside life of the southern tip of Africa.

One of the most interesting places along their route was St. Croix Island off Port Elizabeth, entirely overrun by Antarctic penguins. There the *Calypso* team watched and filmed these fascinating birds as they went about their daily parades, games, and rituals.

From St. Croix, *Calypso* traveled to nearby Bird Island, the home of an incredible number of black-and-white gannets called boobies. In fact, there were so many boobies on this tiny island that in order to walk on its surface, the men from *Calypso* had to push the birds aside with their feet. Interestingly, no matter how crowded the birds were, they always managed to leave a long, narrow strip of open space in their midst—a sort of takeoff runway, for which each bird patiently waited its turn. Like many oceanic soaring birds, boobies have extremely long wings. As a result, they cannot flap their way into flight as smaller birds can but must run along the ground to gather momentum for flying.

On February 29 *Calypso* anchored off Geyser Island, off the South African coast just east of Walker Bay, in order to avoid a storm that was passing through the area. While there, the team took advantage of the delay by filming the sea lions that inhabited the island and surrounding waters. Divers, fearing the self-protective nature and formidable teeth of the sea lions, descended in antishark cages, but they soon left the cages to film in the open sea. The sea lions, curious and attentive, clustered around the cameramen, but scattered at any sudden movement.

Sea lions are intriguing animals, and Captain Cousteau had long wanted an opportunity to study them in depth. His team's experiences with them at Geyser Island, however, proved that such study would require taking an animal on board ship. While Cousteau was aware of the need for detailed information about sea lions, he also recognized that captivity in any form was unnatural—against everything, in other words, that he and his team stood for. Cousteau weighed the issues carefully before deciding that the knowledge to be gained from such close observation was worth the anguish that taking the animals from their homes would cause to his own conscience. He ordered *Calypso* to Seal Island south of Capetown, where sea lions were plentiful and gave his divers instructions to capture two of the creatures but to handle them with the utmost gentleness and consideration.

After a long and frenzied chase, the landing party finally succeeded in netting two sea lions and towing them, in antishark cages, back to *Calypso*. The sea lions, named Pepito and Christobald, were not injured, but seemed more than a bit shaken by their ordeal. Luckily, they quickly adapted to life on board ship, where a special shelter and swimming pool had been installed for them. Soon they became more like pets than prisoners, and even accompanied *Calypso*'s human divers on their jaunts below the surface. Although at first they were kept on harnesses and leashes, Pepito and Christobald soon progressed to the point that they no longer required these restraints, but dove and swam freely with their human companions and then, like them, climbed the ladder on the side of the ship to return to their new home. All in all, the seven months they would remain on *Calypso* would prove to be an extraordinary experience in communal living.

Calypso stopped but twice on this, her fifth, Atlantic crossing, once at St. Helena Island, to which Napoleon had been exiled, and once at Ascension Island. The stay at St. Helena was brief, and notable chiefly for the prodigious amount of fish—food for the two sea lions—caught there by the crew. At Ascension Island the team witnessed daily rocket launchings (rehearsals for the Apollo space shot) being conducted by the American airbase there, and they again spent much time fishing for Pepito and Christobald. The catching of fish was a new activity aboard *Calypso*: the crew had never indulged in it for their own meals, but since the sea lions required some twenty pounds of fresh fish daily, the men of *Calypso* soon learned to be quite proficient at the task.

Except for these two brief layovers, *Calypso* sailed uninterruptedly from Capetown to Natal, Brazil. During the trip, the team were occupied by two main duties: studying and caring for the sea lions and measuring the amount of micrometeorites—dust particles from space that may affect the quantity of marine life—in various parts of the ocean.

Calypso arrived at Natal on April 12, 1968, to the sound of cheers from spectators gathered on the pier for the occasion. Everyone there had heard of the sea lions, and all were anxious to meet these two new celebrities.

From Natal, *Calypso* sailed north to the mouth of the Amazon River, where the normally deep-blue Atlantic took on a greenish hue. Surprisingly, the *Calypso* team came across a school of dolphins there, strange because although some river dolphins do consistently inhabit fresh water, these marine mammals are almost always found in the sea. The ship continued north, making her way to Puerto Rico, where she arrived on April 25.

The team now found that they had some extra time before they had to begin the archaeological excavation program that was the object of this trip to the Caribbean, so they improvised a program of free dives with the sea lions at such islands as Guadeloupe, St. Barthélémy, Barbados, and St. Thomas. They returned to Puerto Rico in late June and sailed from there to the Silver Bank, a sea-level coral reef as large as a typical French province, situated just northeast of the Dominican Republic. The approach to the reef was extremely treacherous. Ever since the sixteenth century, ships had been broken up trying to cross it, spilling their often-precious cargoes all over the coral; but *Calypso*, with her shallow draft, high crow's nest, hot-air balloon, and other advanced navigational aids, ran into no real problem.

From July 15 to September 4, in unremitting heat and with food and fresh water running precariously low, *Calypso*'s team devoted themselves to the excavation of a sunken ship believed to have been the famous seventeenth-century Spanish galleon the *Nuestra Señora de la Concepción*, sunk while transporting fabulous stolen treasures from the New World. The excavation was exhausting work; it involved handling the temperamental air lift, breaking up tons of coral with sledgehammers, and raising heavy rusted cannonballs and pieces of ancient crockery. After almost two months of this grueling work, they discovered to their dismay that the wreck was not the *Nuestra Señora de la Concepción* at all, but merely a simple eighteenth-century merchant vessel—with no treasure in its hold. This was a great disappointment to the entire team, who had held many romantic notions of treasure buried beneath the sea, and they bitterly rechristened the wreck *Nuestra Señora de la Decepción*.

There was no time for self-pity aboard *Calypso*, however, as plans were already being laid for the next project: an expedition to Lake Titicaca, located high in the Andes between Peru and Bolivia. To get to the lake, *Calypso* sailed through the Panama Canal on September 20, then headed south in the Pacific Ocean toward Peru,

Maintenance work on any ship takes nearly every spare moment of a crew's time; here, a *Calypso* sailor stitches cloth around a rope hawser to keep it from fraying.

where Jean-Michel Cousteau, who had made the preparations for this most difficult mission in the Andes, awaited. Along the way the ship dropped anchor at Foca Island off Peru's northern coast, where pelicans, sharks, dolphins, and swordfish were filmed, adding much spectacular and dramatic footage to Cousteau's television series.

Calypso arrived at the port of Matarani, Peru, on October 2, a gray, cloudy day, and intense physical preparation of the equipment began. The minisubs were made lighter for use in freshwater, for example, and all the necessary equipment—minisubs, decompression chamber, aqualungs, cameras, and so forth—were transferred from the ship to the wharf and then to a special train that would carry Cousteau and his team to Lake Titicaca. Three days later, the men took off on the train for the lake, leaving behind a skeleton crew to man *Calypso*.

Lake Titicaca is immense—3,200 square miles—and is, at 12,506 feet, the highest navigable body of water in the world, but it had never before been sounded or explored in depth. The *Calypso* team's purpose here was threefold: to discover the physiological implications of diving at high altitudes; to answer questions about the nature of the lake itself (for example, its depth, geological origins, and the types and origins of its flora and fauna); and to perform archaeological investigations, such as searching for signs of ancient civilizations through buried treasure or ruins, in particular the lengendary golden chain of the Incas.

One of the most remarkable things discovered at the lake was a species of large frogs, at least 20 inches long, unique to the area. These frogs differed from all other frogs in that they lived completely underwater, and would suffocate if they were brought to the surface. They had adapted to underwater life not by developing gills, as had fish, but by breathing through their skins, as if these were membranes, and thus oxygenating their blood. The divers estimated that there were about a billion of these frogs living in the muddy bottom of the lake—in fact, there was hardly a bit of the bottom that did not have its own frog.

During their stay at Lake Titicaca the *Calypso* team encountered a strange tribe, the Urus, a primitive people who lived in huts built on floating islands made of woven rushes. The Uru culture is the most ancient of this Andean region, antedating not only the Incan but the Aymaran, and the Urus are believed to be the last remaining survivors of the original inhabitants of the South American continent.

While the landing party was exploring the region and the depths of Lake Titicaca, *Calypso* was in dry dock in Callao, north of Matarani, where she received some necessary and long-overdue repairs. Once these repairs had been completed, *Calypso* sailed up and down the Peruvian coast on a hydrological program until the time came to pick up the members of the Lake Titicaca expedition back in Matarani, which she did on November 20.

From Matarani, *Calypso* sailed northward, in rough seas and

stormy weather, escorted by sperm whales, swordfish, dolphins, sharks, and flying fish. She arrived at Acapulco on December 7, stayed there for one day, then proceeded on to Cape San Lucas, on the southernmost tip of the Baja California peninsula, where André Laban and Canoë Kientzy made several dives with the minisubs. On December 19 *Calypso* arrived at Long Beach, California, where she spent the Christmas holidays.

After a short Christmas break at Long Beach, *Calypso* set sail for the island of Guadalupe, situated about 180 miles west of the Baja Peninsula. The purpose of this trip was to locate and film colonies of a rather rare but intriguing marine mammal: the elephant seal. Elephant seals are massive, fleshy creatures—males grow to be as long as 20 feet and are as heavy as three tons. They convey the impression of shapelessness and clumsiness on land, but maneuver in the water with a graceful agility that is quite surprising considering their great bulk. In fact, these are amphibious mammals, who

Two sea fleas cross the Andes aboard a flatcar of a Bolivian train on their way to Lake Titicaca.

Left: looking like an invader from space, diver Jean-Claire Riant stands among a family of Uru people, whose huts are built on floating islands made of carefully stacked layers of the rushes indigenous to Lake Titicaca.

can spend most of the year in the water, only needing to venture out onto dry land to mate and bear their young. Beginning in March or April, these giant marine creatures would leave Guadalupe to return to the water, where they would remain for seven months, living a nomadic life and constantly roaming the vast seas without coming to shore even to rest.

Guadalupe Island was chosen as the optimum site for a film about elephant seals because here they came under the complete protection of the Mexican government. Cousteau assumed that there would be a sufficient number of them here to assure some interesting scenes, although no estimate of the animals' population existed at the time. The plan was to put a team ashore to live with the elephant seals and try to get as much footage as possible, both on land and in the water. The logistics for the mission had taken quite some time to work out, for Guadalupe was really no more than a huge, hostile, barren rock with no resources of any kind, not even fresh water. Therefore, in order to sustain themselves, the team had to bring with them a formidable amount of supplies, including vast numbers of plastic water jars and a quantity of live chickens that served dual purposes as egg-layers and meat supply. All these provisions, along with the Zodiacs, diving saucers, aqualungs, and fuel, were loaded onto *Calypso*, which made the crossing from San Diego to Guadalupe Island in one night, December 27.

As soon as they arrived at Guadalupe, the team were struck by the extraordinary number of elephant seals on the island—so many, in fact, that the beaches were black with the animals. Bodies pressed against bodies and large adult elephants trampled babies in their paths. Luckily, the babies, soft and flexible, almost elastic, were not usually hurt, but their screams added to the general sense of pandemonium. Indeed, the noise on the island was almost deafening, and the great variety of sounds—gurgles, snores, screams, cries, and barks of many kinds—were recorded there by *Calypso*'s sound engineer, Guy Jouas.

In addition to filming the creatures, the *Calypso* team made various studies pertaining to the elephants and their surroundings—weighing them, measuring them, and observing at all times their social behavior. From the results of these studies Cousteau concluded that overpopulation had produced an overaggressiveness among the elephants: a simple territory dispute or case of jealousy became a fight for sheer survival under these abnormally close conditions. Indeed, team members had to approach the animals with great caution, for the beasts would suddenly rise up and charge when they sensed another presence.

Calypso left Guadalupe on January 8, 1969, but left behind a team of scientists and cameramen to continue the observation and filming of the elephant seals. These men stayed for three more months, capturing on film the dramas of courting practices, mating, and birth of these remarkable animals, whose grace and beauty underwater belied their ungainliness on land. The men stayed on the

A baby elephant seal lies on the rocks of Guadalupe Island, off Baja California.

island until April, when the new calves had grown to three hundred pounds and were ready to join the adults in a mass departure for the open sea, where they would live until the next winter approached. Then they would again return to this very island to repeat their yearly cycle.

Meanwhile, *Calypso* was busy tracking the migration of a colony of gray whales, called "desert whales," to Scammon Lagoon, in Baja California, where the whales went every three years to give birth to their young. This was no easy trip for *Calypso*, for the entrance to the lagoon was narrow, heavily shoaled, and filled with salt marshes. The ship touched bottom twice, but fortunately the bottom was very muddy in both instances and no serious damage was sustained. Still, it was deemed safest to use the auxiliary crafts for easy maneuverability, so two Zodiacs and two launches went out every day to reconnoiter the area. Unfortunately, these small, gray, rubber craft were often the object of the young calves' curiosity, to the great anxiety of their mothers. On one such occasion, when a mother saw her offspring consorting with a Zodiac, she rushed to it instantly, sending everything—men, cameras, equipment, and the raft itself—high into the air, to come crashing back to the sea. This done, the mother calmly herded her calf back home. She had felt no particular animosity toward the *Calypso* team, they concluded, but like mothers everywhere had simply tried to ensure the safety of her child.

The filming and study of the whales completed, the *Calypso* team turned their attention to the proliferation of sea urchins in the Coronado Islands, southwest of San Diego. Then, beginning on February 10, the ship underwent a major overhaul at San Diego, where she remained for about a month before the men resumed their seagoing duties, which mainly entailed short trips to Guadalupe Island. Their work was interrupted on March 21 by the news that Daniel Cousteau, Captain Cousteau's father, had just died at the age of ninety-two. This deeply saddened the crew members, for Daniel Cousteau had taken a strong interest and an active role in *Calypso*'s destiny; he himself had started diving at age sixty-seven, and had been an able administrator for *Calypso* for many years.

From March 25 to June 3 *Calypso* was at Catalina Island, off the California coast, circling the area in the attempt to film and measure the speed and diving ability of marine creatures such as dolphins, groupers, and various whales, particularly orcas. These so-called "killer whales" are large, handsome black-and-white mammals universally feared for their combination of high intelligence, enormous teeth, and rapacious appetites.

While anchored one night off Catalina, *Calypso*'s siren sounded, signaling that one of her engines had overheated. Before anyone could get below to inspect the damage, however, the electricity went off, shutting off the water pumps and leaving the ship in complete blackness. By the time the emergency lighting system was activated, the men had gone out onto the deck, and what they saw

A diver in the water with an adult elephant seal.

in the renewed light made them forget all about the power problem: the sea, which only moments before had seemed perfectly calm, was now seething with activity. Bernard Chauvelin and Jacques Delcoutère dove immediately into the water to explore the situation more closely, and found themselves surrounded by squid, some just four inches long and others almost a foot in length, some red and others golden, but all writhing and squirming in constant motion. These squid were discovered to be the cause of the power failure, for their gelatinous bodies had pressed against the screens of the water pumps, blocking the water intake used to cool the engines. As soon as the offending squid were removed, however, others took their places, so a special mesh screen was devised on the spot to keep them separated from the intakes; still, crew members had to make periodic squid-removal dives throughout the night.

On June 3 *Calypso* headed north from Southern California toward San Francisco to dock under the Golden Gate Bridge, where she played host to hundreds of interested journalists and other visitors. From there the ship continued on to Seattle, the city where she had been built. Although the actual shipyard in which she had been constructed no longer existed, several engineers and other workers who had had a part in her creation came to see this now-famous little vessel.

It was just a short trip from Seattle to Alaska, where a new program devoted to sea otters and salmon awaited *Calypso*. Unfortunately, the weather was stormy and the sea quite choppy, so the ship was forced to cut her speed and take refuge behind Graham Island before she finally reached Anchorage on June 23.

Alaska posed a unique problem for the *Calypso* team: the scenery—great glacier-cut cliffs, snowcapped peaks—and strong sense of isolation and timelessness were so striking and the daylight at this time of summer solstice was so long-lasting that the cameramen would sometimes forsake sleep and take photographs for twenty-two hours straight, throwing the entire routine into disorder.

The first stop, Kodiak Island, also provided some unexpected problems. Often the *Calypso* team were forced to leave the ship behind and travel the small winding rivers by Zodiac, sometimes climbing steep inclines, dragging their crafts behind them. They were constantly soaked by waterfalls, some of which hit with tremendous force. Still, the island proved to support much intriguing wildlife.

Kodiak Island is in the heart of salmon country, and since this was the mating season, the fish provided the cameramen with many opportunities to capture their antics on film. During mating season, salmon follow an inner drive that impels them to return from the ocean back to the places of their births. The *Calypso* photographers witnessed the valiant efforts of the salmon as they struggled up streams and rivers against strong currents, in their often-futile at-

Opposite: divers brave a freezing waterfall to film salmon's upstream struggles to their spawning grounds. Often the divers would catch the fish as they fell back exhausted from their attempts to leap the falls, and place them in the waters beyond so that they could continue their journey.

Overleaf: Calypso cruises the coast of Alaska.

A sea otter uses its front paws to eat a crab while floating in the water off the coast of Monterey, California.

Opposite, above: Philippe Cousteau and another member of the *Calypso* diving team take to the water to film the antics of a sea otter at Monterey, California.

Opposite, below: a colony of walruses sun themselves at water's edge on Round Island, Alaska.

tempts to return to their birthplaces—a perfect physical representation of the life cycle.

Kodiak Island is also the home of the great brown bears that share its name and are the largest bears in the world. These magnificent animals stalk the edges of lakes and streams on the island, feeding on the salmon, which they scoop up in their great paws.

Another intriguing creature found at Kodiak was the king crab, a formidable crustacean that sometimes grows to be well over three feet in diameter. The Alaskan king crab is the center of commerce on this island—one can hardly travel anywhere along the coast without seeing enormous traps bulging with the struggling giants. Several huge, antiseptic-looking canneries serve the area, but since the waters surrounding Kodiak are exceptionally rich, the commercial exploitation of the king crab has produced no noticeable depletion of the species.

Calypso left Kodiak Island on July 4, and proceeded to the tiny Cherni Islands in the Aleutians, where the program called for the filming of sea otters, fascinating shoreside mammals which, because of their beautiful coats, had once been hunted nearly to extinction. Conservation laws now protected them. The sea otters exhibited many distinctly human characteristics, which greatly endeared them to the men of *Calypso*, who concluded that Alaskan sea otters were among the most engaging animals they had ever encountered.

On July 9 *Calypso* dropped anchor off the island of Unalaska. That same day astronauts Neil Armstrong and Edward Aldrin first set foot on the moon. The *Calypso*'s log noted the fact and pointed out that while two men were exploring the surface of Earth's barren satellite, Raymond Coll was probably the deepest man on Earth, piloting a minisub some 500 feet beneath the ocean.

The next stop on *Calypso*'s Alaskan program was Round Island

Divers explore the wreck of a Japanese fighter plane, a relic of World War II Pacific island warfare.

near the Alaskan mainland in the Bering Sea, where the beaches were literally covered with thousands of walruses. As soon as a human approached, the walruses flopped heavily into the sea, making it difficult for the men to film them. Their fear of humans was quite understandable, however, for walruses have always been hunted mercilessly for their tusks; in fact, several decapitated walruses were found on the island. *Calypso*'s photographers filmed the hulking beasts on land and in the water, where they were both easier to approach and much more graceful. On land, walruses are not immediately engaging animals, with their bloodshot eyes, massive, flabby bodies, and hostile attitudes; those found on Round Island were no exception. Indeed, these particular walruses comprised perhaps the most unattractive group imaginable, made up as it was exclusively of old males too weak to join the traditional yearly migration northward. Round Island appeared to be a sort of haven for the aged during the summer months, although the group here was joined by females and younger males in the winter. As unappealing as the walruses at first seemed, the animals proved themselves capable of displays of affection as well as intelligence, as soon as the *Calypso* team became better acquainted with them: they remained, nevertheless, rather wary.

From Round Island, *Calypso* headed for Bogoslof Island, west of Unalaska and the westernmost point reached in her travels, then returned through the Akutan Strait to the Cherni Islands, periodically stopping along the way to film some fifteen orcas, all of which appeared quite willing to be movie stars. After additional stops at Kodiak and Anchorage for rest, refueling, and servicing, *Calypso* and her crew made their way back the way they had come, south down the west coast of Canada until they reached Seattle on August 29. *Calypso* received a complete overhaul there, at the site of her construction, and remained docked for the next month.

On September 29, the *Calypso* team was ready to proceed on their next mission: the observation and filming of the giant octopus of the Pacific Ocean. This mission would be exciting: giant octopuses were an intriguing challenge to the divers, for these creatures had long been thought to be the fabled sea monsters of maritime lore, and hardly anyone had ever really studied them in detail before. This would be a dangerous mission as well; a short time before, other divers who had met with giant octopuses in these same waters had had a long and bloody battle with the creatures, and barely escaped with their lives. More important, perhaps, specialists from the Seattle aquarium had warned the *Calypso* men about the dangers of the highly venomous bite of these animals. In addition, the waters surrounding Seattle, where the divers would be concentrating their efforts, were so cold even in these early autumn months that divers ran the risk of hypothermia. Still, the divers were anxious to come face to face with these creatures, some of which measured up to 20 feet in diameter yet had the fantastic ability to stretch themselves thin, like rubber, and otherwise change the conformation of their bodies. The octopus's color can also change, taking on a vivid red or almost violet hue when it is brought to the surface or otherwise disturbed.

In pursuit of these animals, *Calypso* took on board a young marine biologist named Joanne Duffy, a specialist in the study of the giant octopus. She taught the *Calypso* divers how to lure the octopuses out of their underwater lairs, how to tame them, and how to make the creatures cooperative for filming. The biologist proved to be a fine actress as well as an accomplished diver, and the film produced of this mission demonstrates her dextrous ability to handle the giant octopuses, some of which weighed more than she did.

Calypso divers on Truk Island inspect enamel plates retrieved from the galley of a sunken Japanese supply ship.

From Seattle, *Calypso* sailed on October 29 to Monterey Bay, south of San Francisco, where she spent a week filming the area's sea otters. The film produced during this stay includes a remarkable sequence of otters capturing sunfish and tearing their fins off, so that the sunfish remain alive but captive, unable to swim away.

On November 10 *Calypso* left Monterey for Long Beach, which would be her temporary home port in the United States for the next month. From Long Beach she made several short trips to the coastal areas around Catalina Island to study the behavior of dolphins, probably the most intelligent and affectionate of all marine mammals. *Calypso* sailed for Cape San Lucas, at the southern tip of Baja California, on December 10, with the cameramen still filming dolphins, as well as pilot whales, swordfish, and barracudas, along the way. *Calypso* herself was in turn filmed from an airplane that had arrived to document her operations.

The *Calypso* team celebrated Christmas at San Benedicto Island, some 250 miles south of Cape San Lucas, where they dove deep into trenches with the two Sea Fleas to capture on film the many sharks swimming about the area. New Year's Day found the team on their way back to San Lucas with everyone at work shooting a sequence of the two minisubs in action.

After taking on fresh supplies, *Calypso* headed next for Clipperton, a French atoll located about 600 miles west of Acapulco, where the sharks were abundant. There the divers ventured into the water only with the protection of the antishark cages. One shark, about twelve feet long, leapt out of the water several times, providing the cameramen with some spectacular footage. A helicopter owned by an American tuna company also added some aerial shots of the program.

Using *Calypso*'s map room as an improvised laboratory, Dr. Bartholomew administers an EKG to a Galápagos iguana.

GIANT TORTOISES AND BOTTOMLESS BLUE HOLES

After a short trip to Puerto Angel on the southern Mexican coast, where the men went off in Zodiacs and launches to search out and film manta rays, *Calypso* plotted a course for the Galápagos Islands. She arrived there on February 1, and landed a team on the island of Hispana to study and film its inhabitants.

Perhaps the most interesting animal found on the islands is the marine iguana, which is found nowhere else in the world. The appearance of the Galápagos iguanas varies from one island to another, for they spend their whole lives on one island, never venturing to cross the waters that separate the various parts of the archipelago. Therefore, each subspecies—there are eight all together—looks radically different. For example, an iguana found on Hood Island might be red and green, on Santa Cruz it could be bronze and green, and on Cristóbal dark gray in color. Furthermore, since there is no interbreeding between iguanas, these creatures are a sort of living fossil, entirely unchanged since prehistoric times. Indeed, the marine iguana, with its dorsal crest, dragon-like head, and quick, staring eyes, seems very much an anachronism—a modern dinosaur left over from the Age of Reptiles. In fact, no one knows the ancestry of the marine iguana or how and when it first arrived at the Galápagos.

Another fascinating animal found on the islands is the giant tortoise for which the archipelago was named—*galápago* is the Spanish word for tortoise. The giant tortoise is now a protected species on

Above and left: unlike the land iguanas of the American continents, the marine iguanas of the Galápagos have developed behavioral and physiological adaptations that enable them to survive both on shore and in the water.

these islands, but its numbers have been drastically reduced since the days of the pirates, when tortoises were quite plentiful. Both pirates and whalers captured the animals frequently and used them as fresh supplies of meat aboard their ships, for tortoises could be kept alive for over a year without either food or water and slaughtered when needed. Like the iguanas, the tortoises of the Galápagos Islands differ radically from island to island. As a general principle, however, those tortoises with very large, domed shells inhabit the more humid islands, while those with smaller shells occupy the more arid ones.

Sea lions were also found on the Galápagos. The ones encountered by the *Calypso* team were particularly friendly, mingling freely with the men on the island. Sea lions also live in harmony with the iguanas, although they appear to like to tease the reptiles, delighting especially in pulling their tails.

All in all, the Galápagos Islands seemed like a paradise to the men of *Calypso*. Here all the inhabitants seemed to live in peace; no animal exhibited fear when approached by man. The archipelago was indeed, perhaps, an animal utopia.

Calypso spent a total of two months at the Galápagos, during which time the team made several spectacular films, turning their cameras on enormous manta rays, colonies of hammerhead sharks, and strange horned, red-lipped batfish. Curiously enough, on this equatorial island group they also found penguins.

During this period the ship had to maneuver among treacherous reefs and through poorly charted waters, and it would have been almost a miracle had there been no mishap. Such was not the case, however, for on March 27, while sailing between Hispana and San Cristóbal, where Captain Cousteau planned to thank the Ecuadorian authorities for their hospitality, *Calypso* crashed into an uncharted underwater rock. The violent impact shook the ship from stem to stern, overturning tables and chairs and throwing several men to the floor. An immediate inspection showed that *Calypso* had been extremely lucky, however, for the only damage was to her false nose; its observation chamber had been completely crushed and had a large "V" cut into it. Fortunately, the false nose had absorbed most of the impact and the ship's hull was still solid. She had not sustained a single leak, just some superficial injuries and minor damage to a few of her instruments.

Still, it was deemed prudent to give the ship a more thorough going-over before she began her next mission: the study of "blue holes" in the Caribbean Sea. To this end she sailed through the Panama Canal and on to New Orleans, where she was put into dry dock. There her false nose and observation chamber were rebuilt, her keel was repaired, and her exterior repainted. These improvements completed, she left New Orleans on April 29 and set a course for Belize in what was then British Honduras. Here the *Calypso* team would explore the most spectacular blue hole in the world, in the heart of a shallow, flat-topped coral reef called Lighthouse Reef.

The object of study was a legendary, deep, circular hole reputed to be bottomless and with perpendicular walls.

Navigation proved very difficult for *Calypso*, for there was a sort of coral plateau surrounding the blue hole. Members of the *Calypso* team had to reconnoiter the area in Zodiacs and launches, then lay out what they considered to be the least perilous route for *Calypso* to follow: a two-mile-long, tortuously winding channel at times only three feet deep and fifteen feet wide. The ship scraped bottom only once, however—luckily at a sandy spot—and no damage was sustained.

The coral ring surrounding the blue hole was itself brown, a living ring made up of coral, sponges, and sea fans. Once inside the ring, *Calypso* was secured with six nylon lines attached to the coral heads on the edges of the ring; looking down from the helicopter rented for the occasion, she seemed like a spider at the center of a web. As soon as the ship was secured, the unpacking of equipment began. In addition to the now-standard gear, such as minisubs, launches, and Zodiacs, there was a new wet submarine—especially

Left: off the coast of Belize, *Calypso* is moored in a blue hole, surrounded by an almost perfect circle of coral.

In dry dock, workers attempt to reconstruct *Calypso*'s false nose observation chamber, crushed in a collision with an underwater rock off the Galápagos Archipelago.

Because of the unusual depth of the blue hole, *Calypso* was forced to anchor with long lines strung to the shallow coral perimeter of the hole. Here, a sea flea is launched as a second waits atop the rear hold panel.

useful for its great speed underwater and for its cargo of breathing equipment—and there were newly modified underwater scooters. The divers could not wait to begin to explore this great blue lake surrounded by coral. They did not know its topography but sensed that it was very deep, and were anxious to see if it was inhabited.

The divers' first impression below the surface of the blue hole was one of size. At a depth of 100 feet the walls surrounding them curved outward and huge stalactites, more than 40 feet in length, hung down everywhere—some straight, some twisted, some extending down past the divers' range of vision. Down 50 more feet they found themselves in an immense stone forest with gigantic limestone pillars and columns and a massive formation of vaults and "chapels" reminiscent of a Gothic cloister. What was now a blue hole in the sea was once a cave—above the surface of the sea, for stalactites can only be formed in the open air. This means that this blue hole had been formed during the end of the last glacial era, before the sea rose to fill the cave. Further exploration with the minisubs showed this blue hole to be 400 feet deep. The bottom was strewn with pieces of limestone, presumably remnants of the cave's broken roof.

Sea monsters figure prominently in the many legends about blue holes, but the *Calypso* team failed to discover any. They did, however, encounter a variety of marine life, including sharks, jacks, and barracuda, and a few coral fish. Since it must have been as difficult for these creatures to penetrate the coral ring surrounding the blue hole as it was for *Calypso*, their presence here was quite puzzling. At a depth of 250 feet the water became motionless and cloudy, despite the tides, and there were no more fish—which indicated that there was no exchange of water between the bottom of the blue hole and the ocean. Any opening that might once have existed had long since been blocked.

Calypso and her men spent about two months studying this blue hole, diving continually, both free-swimming and in the minisubs. Finally, in order to carry the scientific observations as far as possible, Captain Cousteau decided to have one of the stalactites brought above the surface to be studied in detail. Hopefully, the stalactite would help provide the answers to geological questions about the origin of this region.

Bringing the stalactite aboard *Calypso* was no easy task, but eventually a team of divers armed with sturdy nylon-and-steel cables were able to retrieve their "trophy"—a twenty-foot-long, one-ton segment, which was hoisted by Yumbo the crane to *Calypso*'s rear deck. Pieces of this stalactite would later be examined by several prestigious scientific institutions. It was dated as being twelve thousand years old and yielded valuable information on the geological history of the Earth in this area.

Her mission off the coast of Belize accomplished, on July 9, 1970, *Calypso* headed for the Bahama Islands to study the many smaller blue holes found there and to compare and contrast the results of the studies of the two areas.

For the next month and a half the *Calypso* team were occupied with the exploration and study of the area around the seven hundred islands that comprise what are known as the Bahamas, with particular attention to the numerous blue holes that speckle the seascape. With the help of Dr. George Benjamin, a Canadian chemical engineer and perhaps the person most knowledgeable anywhere about this fantastic world, Captain Cousteau and his diving team sought out the most interesting underwater grottoes. Holes and corridors, often with huge stalactites like those found off Belize, were examined and filmed without incident, although the possibility of violent tidal currents sweeping through the corridors and trapping the divers was an ever-present danger.

On August 22, *Calypso* started her trip back to Europe, returning home after almost four years at sea. After a stop for supplies and fuel at St. George in the Bermuda Islands, she hit the open seas and undertook her sixth Atlantic crossing, stopping along the way at São Miguel in the Azores. On the morning of September 3, the mechanics discovered that the port engine was overheating and the stern vibrating. Luckily, the trouble was nothing more serious than a net caught in the propeller, and *Calypso* sped back home without further incident. She finally reached Marseille, after having concluded a journey of over half a million miles, on September 16, 1970. What had been termed "the great expedition" was now complete.

By now the *Calypso* team were used to proceeding without sufficient rest time, so it came as no surprise to them to learn that there was yet another mission to be tackled that year. *Calypso* was allowed less than two weeks in dry dock before she was off again, on September 29. Her mission this time was to carry out an important topographical program off the coast of Sicily for a group of Tunisian and Italian industrialists. This study, dubbed "Opération Tunicile," consisted of gathering information about the area in order to determine the best route and techniques for the laying of underwater pipelines to transport natural gas from Algeria and Tunisia to Sicily and the rest of Italy.

Calypso, under the command of Captain Alain Thibaudeau, set out from Marseille in such abominable weather and rough seas that she was forced to seek refuge in Toulon. On October 5, she continued her voyage, heading for the Strait of Bonifacio and then following the Sardinian coast until she arrived at Trapani on October 7, where she made a rendezvous with two ships that were to accompany her on this program.

The survey was actually begun the next day; echo soundings were plotted from Tunisia to Sicily, photographic profiles were obtained by use of the troika, and irregularities in the sea's floor were explored from the diving saucer. All this was done under very trying working conditions, for *Calypso* was attempting to follow a very precise route in an extremely well-traveled area and passing ships made little effort to deviate from their courses for her, despite her distinctive signals and the fact that she had the right-of-way.

Bad weather continued to plague the mission. *Calypso* and her team had to contend with stormy skies and currents of over six knots, which made necessary some truly acrobatic maneuvers. Nonetheless, the weather did not prevent them from making some thirty contour maps of the area. Profiles were compiled from troika photographs, corings, dredgings, soundings, and current measurements, as well as from firsthand observations made from the diving saucer. One of the most intriguing things discovered in the area was the hulk of an ancient warship, probably of Phoenician origin and remarkably intact. Albert Falco came across it while piloting the saucer in the middle of the Sicilian Channel.

The first week of November was spent in the Strait of Messina, where the team made six dives in the saucer. The weather did not let

Of Dolphins, Icebergs, and Lobsters

Jacques Cousteau and Philippe Sirout pore over charts in *Calypso*'s radio room.

Top: in addition to five pairs of legs—the foremost of which have pincers—the banded coral shrimp wears three pairs of antennae, for feeling its way along the bottom and for stirring up water currents that bring near the plankton it feeds on.

Opposite: within a special triangular net-walled pen suspended from floats, a *Calypso* cameraman films a dolphin near the Strait of Gibraltar.

121

up. If anything, conditions were even more severe than previously and resulted in broken and lost flashes and demolished troikas. Still worse, a chain from one of the ships accompanying *Calypso* got caught in one of her propellers, which caused a great deal of extra work for the crew, who finally managed to free the ship. Despite these many tribulations, the operation was successfully completed. On November 27, *Calypso* returned to Marseille, where other scientific programs awaited her.

During the early part of 1971 *Calypso* was, as usual, on the move, darting here and there in search of new information about the undersea world. A number of short missions dealing with hydrology and seismology in the Mediterranean had occupied her throughout the winter months. In March she set off on a major expedition to study the physiology and behavior of dolphins in their natural environment.

Here the *Calypso* team were forced to compromise. Although Captain Cousteau had a great aversion to the capturing and imprisonment of any creature of the sea, it was obvious that no significant scientific study of dolphins could be made with different subjects spotted at random in different places. Therefore, a method was needed that would allow the dolphins chosen for the study to swim relatively freely in their natural habitat, yet be contained and separated from those dolphins not in the study. The solution of the team was by no means ideal, but it did provide the opportunity for close observation. They rigged a sort of floating dolphin pen, a net 40 feet deep and suspended from a string of air bags. Once the dolphins were captured and confined in the net, they were documented and recorded as quickly as possible, then released. They were never kept in captivity more than two or three days, and some were set free after just a few hours. In this way the subject was never really isolated from his fellows, for he could still hear and see them through the netting—a highly important fact since dolphins are such

Two dolphins cavort in the net pen set up for the Gibraltar dolphin study.

In another study, Captain Cousteau and Dolly, a dolphin trained in reconnaissance by the U.S. Navy, swim together in the canal waters of the Florida Keys.

Albert Falco and Yves Omer mount an underwater camera on the back of a dolphin aboard *Calypso*.

genuinely social beings. Indeed, this was proven by putting two dolphins in the pen together; a noticeable change in their behavior occurred, and the dolphins developed an obvious bond of affection, rubbing against each other and swimming side by side.

Other experiments performed at this time included strapping a radio transmitter and an underwater camera to the back of a dolphin. In this way the *Calypso* team could pick up and record the sounds made by the dolphin as well as track his progress through the waters. At the same time, the camera and transmitter produced a film of the ocean world from a dolphin's point of view—a unique vantage point. The photography on this mission was supervised by Jacques Renoir, great-grandson of the famous painter and a grandson of the noted movie director.

The study of dolphins was interrupted in the late spring so that *Calypso* could participate in a series of seismology programs off the coast of France. She returned to port in early July, and then on the sixth of the month, under the direction of Captain Alain Bougaran, left again, this time to shoot a film on the precious red coral found in the Mediterranean off Corsica. For the next month Albert Falco and the rest of the cameramen devoted themselves entirely to capturing on film the dangerous exploits of the coral fishermen off the coast of

Attempting evasive deception, an octopus changes color to match its environment.

Below: a bright starfish contrasts sharply with a highly poisonous scorpion fish, whose configuration and color blend with the sea bottom.

Opposite: off Riou Island, Jean-Pierre Genest, framed by a hedge of sea fans, observes a Mediterranean octopus.

Propriano and in the Strait of Bonifacio. They witnessed divers taking incredible risks by using only conventional air tanks to swim down as far as 330 feet in order to gather coral from deep caves in the area. The rising price paid for coral by jewelers had created an unprecedented greed among these divers and many fatal accidents resulted from their lack of caution. *Calypso*'s team used the Galeazzi chamber and helium-oxygen mixtures to ensure safety at such great depths.

By late summer, Captain Cousteau and his top aides became monopolized by unforeseen administrative problems, and *Calypso* remained idle for some months. As winter approached, however, she sailed down along the western coast of Africa to Mauritania, where her team could once again devote themselves to getting to know dolphins better. Dolphins hold an important place in the legends and lives of one of the peoples of Mauritania, the Imragen. The Imragen regard the dolphins as special friends. It is considered a sin to kill a dolphin here and the dolphins—so the legend goes—in exchange for this generosity reward the Imragen by pushing fish into their nets.

Mauritania is a study in contrasts. Here the Sahara Desert meets the Atlantic Ocean, arid dunes meet the sea, and the people spend the greater part of their lives in the search for water. But the Imragen, a small group of people divided among only a few huts, have turned their backs on the inhospitable land and placed all their hope for survival in the sea. The abundance of fish in the area is absolutely incredible. From September until February, when vast schools of mullet pass by here on their migratory routes, this is one of the richest areas of marine life in the world.

The men from *Calypso* hoped to witness the legendary spectacle of dolphin helping man in his pursuit of fish. What they saw was indeed remarkable. As a school of mullets approached, the Imragen fishermen set their nets and began beating the water with sticks; a group of dolphins obediently appeared, pushing the mullets before them. A frenzy ensued as dolphins, mullets, and men mingled in the water—the dolphins literally herding the fish into the men's nets. When the majority of the mullets had been confined in the nets, the dolphins headed back out to sea, only to return again a few hours later as another school of mullets passed by. After the day's astonishing spectacle had ended, the *Calypso* team joined the Imragen in celebration, expressing their joy and gratitude to the dolphins by dancing and singing until well into the night.

Guy Jouas and Dr. François listen to sound tapes recorded during the dolphin study near the Strait of Gibraltar.

ANTARCTICA BECKONS

For *Calypso*, most of the year 1972 was spent in dry dock in France, where she remained inactive for many months while undergoing maintenance and repair work. For Captain Cousteau and the rest of the *Calypso* team, however, this was not simply a period of rest between missions. Instead, they put their time to good use,

From the bridge, Captain Cousteau supervises the lift-off of the helicopter from the pad installed on *Calypso*'s bow.

preparing for one of the most difficult and potentially dangerous expeditions in *Calypso*'s entire history: the exploration of the frozen continent of Antarctica.

Although *Calypso* and her team had spent several months in Alaska in 1969, they had never ventured into the polar regions. After much pondering and many discussions with both friends and advisers, Captain Cousteau decided to remedy that situation by going to Antarctica to explore the icy depths of the ocean surrounding this largely unknown, desolate land. Cousteau intended this to be a comprehensive expedition, with information garnered from free dives and descents in the minisubs as well as conventional instrumentation—in other words, the most complete study possible of this forbidding habitat. He worried that *Calypso*, although a sturdy little ship, would be too small and otherwise not built to withstand the arduous journey. Still, she was the only vessel at his disposal and had always proved herself quite trustworthy. So he placed his faith in her and proceeded to organize and set into action a major expedition to the area.

After much repair and renovation, including a helicopter pad that was installed on her bow and substantially altered her silhouette, *Calypso* was ready to sail. She set off from Monaco on September 29, 1972, following a reception aboard ship in which many friends and supporters—including Prince Rainier and his son,

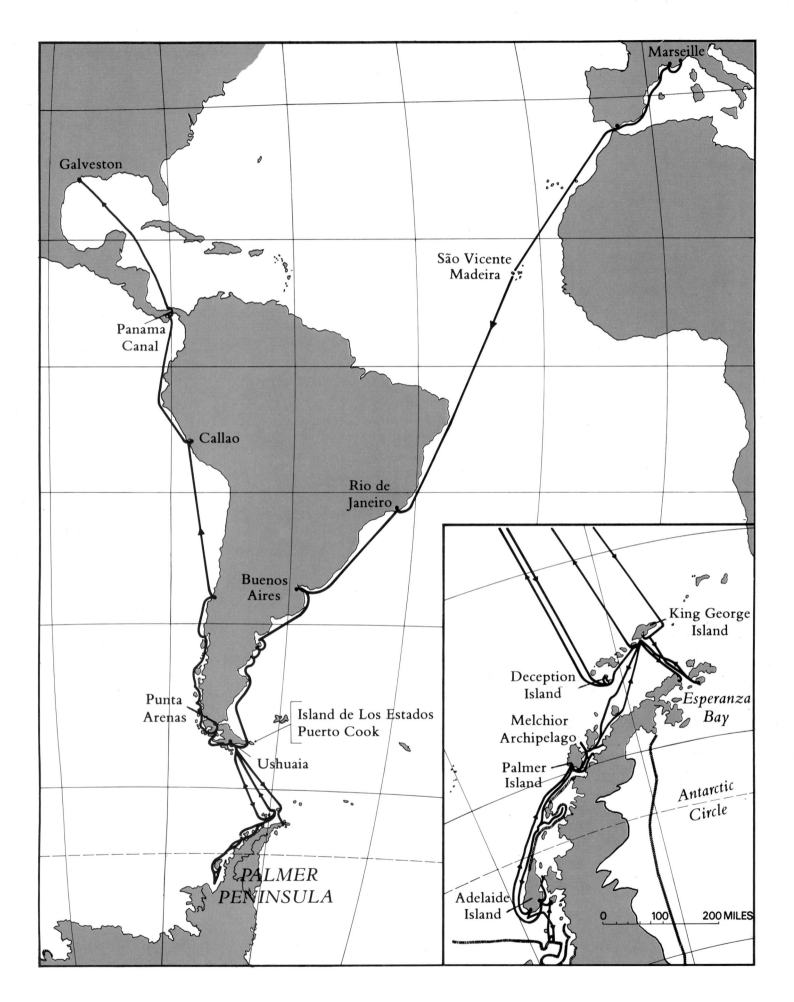

Marseille

Galveston

Panama
Canal

São Vicente
Madeira

Callao

Rio de
Janeiro

Buenos
Aires

Punta
Arenas

Island de Los Estados
Puerto Cook

Ushuaia

PALMER
PENINSULA

King George
Island

Deception
Island

Melchior
Archipelago

Palmer
Island

Esperanza
Bay

Antarctic
Circle

Adelaide
Island

0   100   200 MILES

Prince Albert—gathered to extend their best wishes for her success. The ship, loaded to capacity with all sorts of new equipment, immediately encountered bad weather just off the coast of France. In fact, the weather was so ominous that the helicopter pad was taken down and put into storage three days after departure, for fear that it might prove dangerous in the high seas.

On October 3 *Calypso* pulled into port at Gibraltar to have her radar system, damaged by the storms, repaired. She remained there until the next day, when she started her seventh Atlantic crossing.

Calypso was accompanied on her voyage across the ocean by several dolphins who played continually in front of her prow as she headed for her first stop, Porto Grande, the finest natural harbor of the Cape Verde Islands. After a brief layover there, she continued on her way across the Atlantic, reaching Rio de Janeiro on October 23.

From Rio, *Calypso* made her way farther south, to Buenos Aires, where she was met by a team of American scientists from NASA, who installed a battery of instruments that brought her into the space age. Among the types of equipment was a receiver for the direct reception of photographic images in either infrared or visible light from satellites passing overhead; the NASA team also set up a new system of radio communication and direct-image transmission that employed as a relay station an existing stationary telecommunication satellite. *Calypso* was the first ship in history to be equipped with such modern telecommunication technology.

Above: a mother humpback whale keeps her calf near the surface of the water so she can help it upward for air; she appears unconcerned by the diver holding onto her fin.

Left: off the coast of Patagonia, Argentina, *Calypso* team members in Zodiacs film a right whale surfacing, its double spout actively blowing.

Next on *Calypso*'s agenda after Buenos Aires were stops at Golfo San José and Golfo Nuevo along the coast of central Argentina. Here her purpose was to pick up a camera team, led by Philippe Cousteau, which had been sent there previously to study and film, if possible, the mating rituals of the right whale. Right whales had become extremely rare, having been brought to the brink of extinction at the turn of the century. Their numbers, however, had been on the upswing since whaling became an internationally regulated industry in 1946. Philippe and his crew were, indeed, able to spot some of these rare but handsome cetaceans and capture on film a few minutes of their actual mating before the whales became aware of the cameramen's presence and fled.

At Puerto Madryn, on the western shore of Golfo Nuevo, a new addition to the ship's complement—a Hughes 300 helicopter—was brought on board. The helicopter, boldly painted yellow and black to match *Calypso*'s other equipment, suited the team's needs perfectly: it was small enough to be kept in *Calypso*'s hold, yet could remain airborne for up to five hours, enabling its pilot and one passenger to reconnoiter a large area as well as providing an aerial vantage for films of her operations. The helicopter would prove invaluable; *Calypso* would never again sail without it.

On November 28 *Calypso* arrived at Ushuaia, the chief settlement of the Argentine province Tierra del Fuego and the southernmost town in the world. Here she met up with the *Bahía Aguirre*, an Argentine ship that was to serve as *Calypso*'s supply vessel throughout her Antarctic expedition. The Argentine vessel would also help transport some of *Calypso*'s equipment, such as the diving saucer, which, during the treacherous crossing of Drake Passage between Cape Horn and the great southern continent of Antarctica, was put in the *Bahía Aguirre*'s hold instead of occupying its usual place on *Calypso*'s quarterdeck.

On December 5, with snow falling, *Calypso* set out in seas whipped by cold, violent winds and clogged with drifting ice and mountainous icebergs. Her destination was King George Island, one of the South Shetland Islands just off the point of Graham Land, the northern tip of Antarctica and a part of the Palmer peninsula, about 250 miles north of the Antarctic Circle. After a rough four-day crossing, during which she received several weather maps a day via satellite, the ship reached the island. Immediately upon arrival, some of the crew set up the helicopter pad on deck and lifted the aircraft from the ship's hold. Those of the team not occupied with the copter went ashore and spent some time reconstructing the skeleton of a large blue whale, a victim of the massive hunts carried out in this area in the late nineteenth and early twentieth centuries. The bones were sorted out from the many thousands found on the island, remnants of thousands of whales that had been butchered there in the days before the introduction of the huge factory ships that began to process whales at sea in the mid-1920s.

From King George Island the *Calypso* team sailed to Deception Island, still farther south, arriving on December 19. Deception Island is a truly fascinating place: a circular island, about nine nautical miles in diameter, enclosing at its center a landlocked harbor. The *Calypso* cameramen found many interesting subjects here: seals dotted the icy beaches, and enormous colonies of three different breeds of penguins—hundreds of thousands of the birds in all—were in the process of building nests. In addition, the team photographed several volcanoes, at least one of which was still smoking through the ice that covered it. From one caldera, an immense volcanic crater with a diameter many times the size of an ordinary volcanic vent, is-

Right: Captain Cousteau keeps a careful watch as *Calypso* advances slowly into a natural glacial harbor in Antarctica.

Below: hundreds of Magellan penguins dot the landscape of this Tierra del Fuego rookery, or breeding ground. The crater-like nests are maintained by successive generations of penguins.

130

Opposite: the red suits of the *Calypso* team provide a striking contrast to the stark white of a mountainous Antarctic iceberg.

In a moment of frivolity, diver Dominique Sumian overinflates his insulated dry suit, especially designed for frigid waters.

sued clouds of steam from hot-water springs located deep within it.

Soon the *Calypso* team settled down to a routine, and life in Antarctica became familiar. Daily dives were made both with and without the diving saucer; the saucer went five times into the caldera in the first few days at Deception Island and free-swimming divers equipped with new, completely waterproof suits discovered to their delight that they could remain in the icy waters for up to forty minutes at a stretch without suffering adverse effects from the cold.

Time passed relatively uneventfully for the men of *Calypso*, whose scientific routine these first few weeks in Antarctica was broken only by the celebration of Christmas. Their merriment was cut short, unfortunately, as tragedy struck the expedition. On the morning of December 28, Michel Laval, *Calypso*'s first mate, was on Deception Island to study the properties of ice at the island's higher altitudes. He went to meet the helicopter on one of its frequent shuttles between the ship and the island. As the copter landed on the ice, Laval either tripped or slipped and was struck by the helicopter's tail propeller. He was killed instantly.

Shock and grief pervaded the team, and *Calypso* had the sad task of taking Laval's body back to Ushuaia, where it was put aboard a plane to Paris, accompanied by Captain Cousteau. The program in Antarctica continued, however, for eleven men and much equipment had been left on Deception Island at Whalers' Bay, formerly a military base owned by the United Kingdom.

Calypso returned to Deception Island in early January 1973 to pick up the team she had left behind. From there she sailed still farther south through the Strait of Guerlache and on to the Melchior Archipelago, where she was met by the *Bahía Aguirre*. On Janu-

Swimming through a tunnel of ice, a diver explores the shifting crevasses beneath an iceberg. Small quakes can be heard and the icy walls are rippled from the erosive action of the salt water.

ary 15 she steered a course for Palmer Station, a United States military base on the island of Anvers. Continuing south, she ran into bad weather; temperatures dropped, the amount of ice in the water built up, snow fell, and the visibility dropped to almost zero. The ship drifted slowly, often having to change course—sometimes very suddenly—to avoid hitting packs of ice. Navigation was possible only with the help of the helicopter, which was sent ahead to survey the area and radio information back to the ship. As *Calypso* progressed farther southward, the weather conditions worsened, and the icebergs became even greater in size. The *Calypso* team crossed the Antarctic Circle on January 19, and encountered a huge iceberg, 230 feet high and shaped like a giant sphinx, which they could not resist exploring.

On January 20 they stopped at their first port within the Antarctic Circle, an English base at Adelaide Island. The next day the weather turned unusually fine and clear, so *Calypso* ventured out into a large ice pack south of Marguerite Bay, where the scenery was breathtakingly beautiful, stark and clean; it provided the film crew with some spectacular footage. Philippe Cousteau took advantage of the calm and ascended several times in his hot-air balloon, which gave him a magnificent vantage point from which to view the scene. That evening, *Calypso* berthed beside an iceberg, and divers descended in the saucer to depths of almost 800 feet along the wall of ice.

Exploration continued around Pourquoi Pas Island, in Ryder Bay and Laubeuf Bay. Then, after a difficult passage through the narrow Gunnel Channel, which separates Hansen Island from Arrowsmith Peninsula, *Calypso* again berthed, this time alongside an ice barrier in Hanusse Bay. Here a long underwater sequence was filmed on the seals of the area.

As soon as the weather turned bad again, *Calypso* turned around and headed back to Palmer Station, battling blizzard conditions along the way. Again the helicopter was put into use to help guide the ship on her way between ice floes. From Palmer Station, *Calypso* visited the Strait of Guerlache, where dives in the saucer at 820 feet revealed undersea life forms unique to this area, and never before seen by man.

On February 8 *Calypso* reached Esperanza Bay, an Argentine base at the entrance to the Weddell Sea. The wind was raging as a storm of unusual intensity whipped across the bay; although the ship was at anchor, she was forced to keep her engines running to hold position against the wind and to avoid running into the huge ice floes that were streaking past the hull. In spite of this, one of the huge masses struck *Calypso* on the stern, ripping some planks—fortunately above the waterline—and also hitting the port propeller, breaking its shaft. The blizzard continued for two days and the gale winds reached 100 miles per hour. *Calypso* was covered by a veritable shell of ice that weighed her down to the point of endangering her stability in the water. Thus weighted, running on a single twisted propeller, and entirely dependent upon radar to guide her in zero visibility, *Calypso* maneuvered for three days and three nights in the attempt to escape disaster.

Finally the wind died down a bit on February 11. After dropping anchor in Esperanza Bay, the crew set about clearing the ship of her carapace of ice. As much as possible, the damage done to *Calypso* by the storm was repaired. The planks were put back in place, and divers went underwater to secure the propeller shaft so that it would not jam the rudder. All the alternatives were explored in order to determine the best way for the ship to get out of the bay and into a safe haven.

Calypso was now ready to begin her return journey. But the damage she had suffered was so extensive that American and Chilean ships nearby were prompted to propose to Captain Cousteau that they tow his vessel back to Argentina. Nevertheless, this was not the first time *Calypso* had been damaged, and Cousteau decided that she must be able to get out of difficult situations under her own power. Besides, she was the only ship in the world equipped with NASA's ultramodern satellite communication systems. If *she* could not get back, what ship could?

Calypso set a course northward and sailed toward Ushuaia, preceded by the *San Martín*, an Argentine icebreaker. She arrived at King George Island on February 14 and waited there for two days, until the NASA weather information indicated that it was the optimum time to cross the dreaded Drake Passage.

On February 16 *Calypso* sailed, on the power of a single engine, for Ushuaia, accompanied in her travels by a Chilean naval vessel called the *Yelcho*. Four days after her departure she arrived safely in port, to the relief of all—not only the members of the *Calypso* team, but all those supporters in France, Chile, Argentina, the United States, and throughout the rest of the world who had waited anxiously during her passage to learn of her fate.

From Ushuaia, *Calypso* sailed through the Strait of Magellan to the Pacific Ocean and on to Punta Arenas, the southernmost Chilean port and a naval arsenal. There her broken shaft was replaced, the propeller more professionally repaired than had been possible underwater, and the gaping hole in her hull patched up. While the ship was laid up undergoing these repairs, the members of her team were busy on a new project: a study of the land and marine life of the Patagonia channels and of the life-style of the Indians who continued to inhabit the surrounding regions.

Following her repairs, *Calypso* was floated on March 8 and resumed her travels throughout the channels, where her team occupied themselves with filming aspects of the profoundly difficult lives of the twenty-seven surviving members of the Kawashkar tribe—the last people in the world to live a nomadic life on the sea. Then she returned to Punta Arenas, and from there traveled northward to Puerto Montt, where Captain Cousteau and ten other members of the *Calypso* team disembarked. The diving saucer was also put ashore there, to be sent back to France by freighter.

From Chile she set out for Galveston, Texas—which had been selected as the site for her annual careening—making a single stop along the way at Callao, one of the finest harbors on South America's Pacific coast and the major port of Peru. She arrived at Galveston, after sailing through the Panama Canal, on May 9, 1973. She would later be met there by Jean-Marie France, her chief engineer, who would be responsible for overhauling *Calypso*.

The engineer's task would indeed be a difficult one, for the ship certainly bore the scars of her arduous Antarctic expedition and nearly a year's neglect. But the mission to the bottom of the earth had not been without reward for *Calypso* and her team. The cruise—which had included its share of dramatic moments—had resulted in four films for television and one full-length movie, all of which dramatically testify to the previously little-known magnificence and special beauty of the marine life of Antarctica.

CARIBBEAN SURPRISES

After spending over a year in dry dock, during which time she was thoroughly repaired and renovated following her arduous experiences in the Antarctic, *Calypso* was finally ready to leave the Todd Naval Shipyards in Galveston to resume her scientific explorations. A new program was in the works for the Caribbean and the Gulf of Mexico and Captain Cousteau and the members of his team were all eager to get started.

Calypso sailed on October 17, 1974, for Mujeres Island, just off

In the waters off the northeastern tip of Yucatán, a diver swims
beneath a manta ray with a wing span of nearly 26 feet.

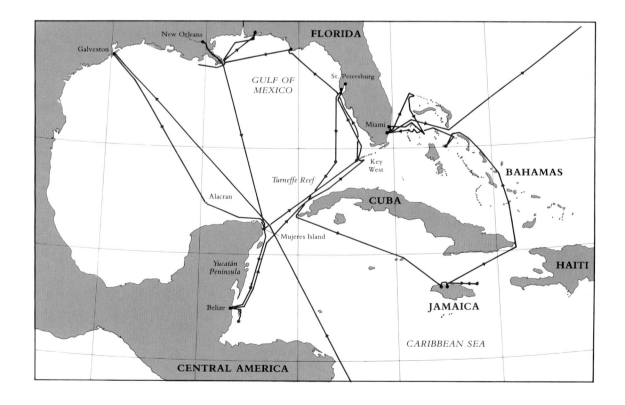

137

the northeastern tip of the Yucatán Peninsula, where the objective would be to locate and observe the sharks that were reported to spend their time sleeping in waters off the coast. Before reaching Mujeres, however, she broke down (due to her long period of inactivity or perhaps to some negligence at the shipyards) and had to make a three-day layover at the Alacrán Reef, eighty miles north of the Yucatán coast. The *Calypso* team, never inclined to idleness, took advantage of the delay and used the time to inspect the many shipwrecks littering the windward edge of the reef.

On October 23 they set out again for Mujeres Island, arriving there three days later. They first made a systematic study of the hydrological conditions in the many caves that existed under the water surrounding Mujeres Island and discovered, at 82-foot depths, ocean-dwelling bull sharks and lemon sharks that were, indeed, motionless in the water and apparently asleep. This was quite a revelation to the *Calypso* team, for it had long been thought that these

Hundreds of thousands of schooling grunts move in harmony as if a single organism.

Calypso's PBY Catalina amphibious aircraft floats serenely beside a landmark of Isla Isabella.

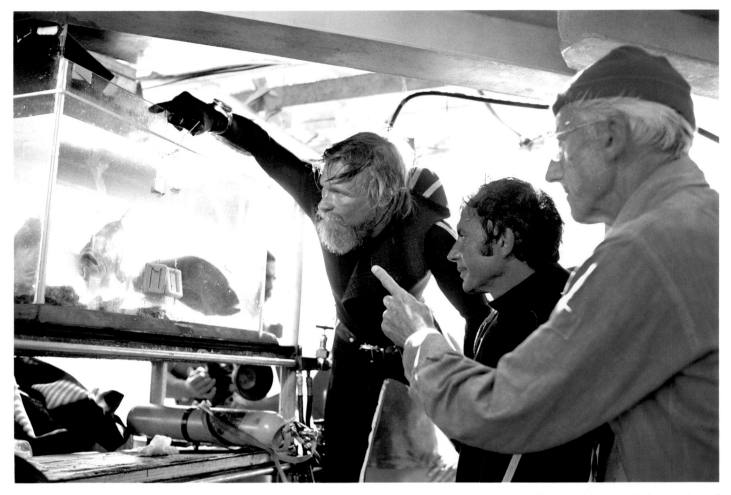

In an effort to photograph the mating of a pair of groupers captured in the waters off Belize, the Cousteau team has set up a saltwater aquarium aboard *Calypso*. Here, Bernard Delemotte, Albert Falco, and Captain Cousteau observe the amorous pair.

the searching proved to be in vain, however, and *Calypso* soon moved on, traveling a bit farther north each time and finally reaching Cape Catoche. Here they did find a few small groups of spiny lobsters, but these showed no signs of imminent migration. Instead, they remained couched in their caverns, seemingly waiting just like *Calypso* for some mysterious omen signaling that the migration was to begin.

The ship had to abandon the area temporarily on January 3, 1975, because the wind (blowing from the wrong direction, according to Válvula) had made the water so turbid that filming became impossible. The wind kept getting stronger and stronger, until finally, on January 9, *Calypso* was forced to return to shore for shelter. She shuttled back to sea as soon as the wind abated, but the team found to their dismay that they had missed the spiny lobsters' migration altogether! On their way back to the open sea they came across fishing boats filled to their gunwales with lobsters, which had apparently swum right by the *Calypso* divers by the thousands. The water had been too murky. *Calypso* had spent more than a month waiting for the migration, but it had taken only a single day of 30-knot winds to render her patience futile.

The stay at Mujeres Island was not a total waste of time. Before leaving the area, the *Calypso* team was witness to and captured on film a truly remarkable spectacle: a huge shoal of fish packed together so tightly that from the helicopter it looked like an oil slick. Only from close up could individual life forms be distinguished.

After leaving the Mujeres Island area, *Calypso* called at Belize, the major port of British Honduras, where she explored the Glover Reef, and then spent two weeks at the edge of the reefs south of Belize, where thousands of groupers gathered each year to spawn. The local fishermen took advantage of the presence of so many of these fish by going out in dugouts and outboards to catch on hooks and lines hundreds of large, live groupers that would later be put into bamboo fishponds, then retrieved and killed as needed.

Next on *Calypso*'s agenda was Turneffe Reef, thirty miles out from Belize City, the site of magnificent coral landscapes. The cameramen spent over a week there, filming the beautifully colored coral and other marvelous marine life, notably the giant sponges.

It was then decided that Calypso should put in at St. Petersburg, on the western coast of Florida, for a well-deserved four-month rest for both ship and crew. At the same time, Jean-Pierre Le Flohic, a former French merchant marine captain, was named commander of *Calypso*. He took advantage of the four months to familiarize himself with every aspect of the ship. On July 20, after she had been completely overhauled and a new helicopter loaded, he officially took command of *Calypso* and prepared for her next mission.

Calypso's first journey under Le Flohic's command took her to Montego Bay, Jamaica, where she arrived on July 24. From their base, *Calypso*'s team conducted an exhaustive study of the magnificent coral reefs off the island's northern coast. While in the area they decided to explore the wreck of a sunken ship, which maps indicated was situated near the Formigas Shoal, about one hundred miles to the east. From the helicopter, piloted by Bob Braunbech, the wreck was located easily because of the extraordinary clarity of the water, and *Calypso* was guided to the area by radio.

Working closely with marine biologists from a station at Jamaica's Discovery Bay, *Calypso*'s divers explored and filmed the spectacular coral cliffs extending from 30 to 650, sometimes even 1,000, feet below the surface. With the aid of a time-lapse photographic device improvised aboard the ship, they were able to film the unusual phenomenon of "cannibalism" that occurs among certain species of coral. They also captured on film unusual flower-shaped coral formations found at depths of up to 200 feet, and other colonies of coral living about 250 feet below the surface.

From the coral reefs north of Jamaica, *Calypso* skirted around Cuba and headed toward the Bahamas, which had been chosen as the site for a new series of experiments planned in collaboration with NASA. Immediately upon arrival at Nassau, on August 24, she participated in experiments designed to test the use of two satellites,

A Nassau grouper, photographed off Belize in the Caribbean, changes color for camouflage. During their spawning season, groupers undergo especially dramatic color changes.

LANSAT 1 and LANSAT 2, to verify or update marine maps of the shallow areas of the surrounding seas—those areas of which accurate knowledge is most essential to safe navigation. Thus, while the satellites were photographing three well-defined areas of the huge Bahamian reefs from the sky, *Calypso* was taking soundings of the bottom and measuring the transparency and reflective properties of the water and of diverse marine sediments. This program employed all *Calypso*'s resources—sonar, helicopter, small craft, and dives, both free-swimming and with the saucers—to make the maps as accurate as possible. The results, which would be analyzed the following year, proved that satellites can indeed be very useful in keeping navigational charts up to date, particularly at water depths of under one hundred feet.

On September 8 *Calypso* returned to Miami, the expedition at a close. After receiving some minor repairs, she began the long journey home—her eighth Atlantic crossing. She reached Marseille on October 7, underwent a complete overhaul there, and finally, on October 24, once more as good as new, returned to Monaco, where plans were already being laid for her next expedition.

Opposite: a blue chromis, a type of damselfish that feeds on plankton, swims among tube sponges and coral.

Before returning to Europe, *Calypso* receives a necessary overhaul in dry dock.

Above: Captain Cousteau films his divers observing the numerous organisms that live symbiotically in a coral reef.

Left: like corals everywhere, this Caribbean variety forms a densely packed community. The green substance at the base of the corals is algae, which the corals farm as a food source.

Opposite: a diver explores the wreck of a vessel recently sunk in the Caribbean near Jamaica, where the treacherous shallow coral reefs have taken their toll on many ships for centuries.

B y the time *Calypso* reached her home port in Monaco the year was almost over, but for the members of her team the autumn of 1975 signaled the beginning of yet another major undertaking. For the next twelve months *Calypso* would be involved in an underwater archaeology program in the Aegean Sea. Cousteau and his divers would search there for new traces of vanished civilizations: submerged cities, ancient harbors, shipwrecks from all ages.

Before leaving for the Aegean, however, Captain Cousteau held a press conference in *Calypso*'s mess—a room much too small to contain the large number of media representatives who gathered. Cousteau explained the scientific objectives of the forthcoming program, expounded on the many hidden wonders of the sea, and announced that the funds for the expedition were all provided either by film contracts or by the newly formed, nonprofit American foundation called the Cousteau Society. From that time on, the funding for all missions and the operation of *Calypso*—including all her equipment, such as the helicopter, the diving saucers, and other instruments still in the developmental stage—was to be entrusted entirely to the Cousteau Society.

The next day, October 31, 1975, *Calypso* sailed from Monaco under the direction of Captain Pierre Mahé. Following the west coast of Corsica and sailing through the Straits of Bonifacio and Messina and the Channel of Corinth, she arrived five days later at her immediate destination, Marina Zea, near Piraeus, on Greece's southern coast, five miles south of Athens.

The task confronting *Calypso*—finding sunken remains of ancient civilizations—was immense. Fortunately, this ship was exceptionally well equipped for the job. On board was a brand-new "side-scan" sonar system developed by Professor Harold Edgerton, an old and valued friend of *Calypso*. Edgerton himself also came on board to help in the operation of his invention, which would enable *Calypso* to detect objects on the bottom of the sea from up to 500 yards away on each side of the ship.

Locating underwater archaeological sites requires, of course, more than technologically advanced equipment. Undersea discoveries come about largely due to the combination of logical reasoning and patient research—making hundreds of inquiries among the native fishermen, for example, or rummaging through historical

A Remembrance of Things Past

Captain Cousteau reads navigational data to determine *Calypso*'s position for underwater recovery work.

Top: dwarfed by thirty-foot-tall sculptures carved of volcanic rock, Captain Cousteau learns something of the mysterious culture of Easter Island, off the southern Pacific coast of South America.

Opposite: an ancient Mediterranean pottery jar bears the marks of long immersion in the sea—a coating of organic growth.

archives for accounts of shipwrecks, battles, or natural catastrophes. Fortunately for the *Calypso* team, they had the helicopter, which shuttled between the islands, the ship, and the neighboring ports to facilitate communication.

Calypso's first excursions in the Aegean Sea took place off the coast of Zea; from there she made her way southeast to Patroklou, then on to Kea Island, where the crew began systematic plottings of the area. After five days of traversing the seas, and making soundings, scans, and plottings, on November 11 they located a large sunken vessel at a depth of over 370 feet. It was quickly determined to be the wreck of the *Britannic*, sister ship of the *Titanic*. The largest vessel afloat at the time, the *Britannic*, serving as a hospital ship during World War I, was sunk under mysterious circumstances in 1916. The *Calypso* team did not then have ample time to examine the wreck closely, but they marked their maps with the intention of returning to the site soon to explore the mystery.

First, though, they yearned to find remains of a ship much more ancient. They spent the entire next month seeking out traces of the vessel that must have carried the magnificent bronze statues that had been found near Cape Artemision at the turn of the century. The search for this ship took them all around Greece, but only some pieces of pottery and one bronze vase were discovered. Although these artifacts were from the same era as the statues found off the coast of Cape Artemision, there was no way to prove that they were from the same ship; and it was the ship they had been investigating, after all.

Next on *Calypso*'s agenda was a thorough probing of the Bay of Pylos, site of the famous Battle of Navarino during the Greek war for independence. The explorations here took another month of *Calypso*'s time, since a considerable amount of debris on the muddy bottom registered on the side-scan sonar and demanded investigation by the divers.

From the Bay of Pylos, *Calypso* proceeded to Herakleion, in Crete, and started a careful investigation of Dia, an uninhabited island off Crete's northern coast. Here they located four ancient shipwrecks, as well as a great many huge Venetian anchors. Just as *Calypso* was about to leave the area for new grounds, Albert Falco made an important archaeological discovery: he identified, in Dia's Bay of St. George, a submerged harbor, probably over three thousand years old. Excavation of this exciting site was planned, but it was determined that special equipment, not at *Calypso*'s immediate disposal, would be needed. While that equipment was being prepared, *Calypso* went on to the southeast coast of Crete, where another important discovery was made: a wall of Minoan pottery was found close to the island of Pina, at a depth of about 100 feet.

From February 6 to March 3, *Calypso* was in the port of Marina Zea. While some maintenance work was done, a portion of the crew was allowed to return to France for a much-needed vaca-

Opposite: Calypso's helicopter proved invaluable during the Aegean operations, facilitating communications and transport between the ship, islands, and ports.

Below: paying its respects, *Calypso* cruises past the Temple of Poseidon, Greek god of the sea, on Cape Sounium at the southern tip of the Attic peninsula below Athens.

tion. The other team members occupied themselves with research, investigating additional possible sites for archaeological exploration.

On March 24 Captain Yves Gourlawen took command of the ship, now carrying an improvised pontoon, a new air compressor, and an underwater air lift, all on board for later installation in the Bay of St. George in Dia. During the months of April and May, *Calypso* made numerous shuttles between Herakleion and Dia. Hundreds of amphoras, pottery, copper and silver plates, marbles, and wood from the hulls of the four wrecks were salvaged and photographed, their locations carefully marked on maps.

During this work, *Calypso* made a foray to a famous sunken ship at Andikithira. The site was rather difficult to identify but the *Calypso* team located it and carefully photographed the area. Cap-

tain Cousteau compiled the pictures of the bottom of the sea into a photomosaic that perfectly represented the site.

Once this week-long diversion had been completed, *Calypso* was again devoted full-time to excavation at the Dia site. After two months of efforts, a team onshore completed the systematic underwater excavation of an area measuring ten yards by ten yards, located close to the sunken jetty of the harbor. The air lift allowed the divers to reach all the way to the base rock, ten feet under the muddy bottom of the sea. In the lower layers of sediment they found Minoan pottery—indications that the harbor was already in use as long ago as 2000 B.C.

152

Thira (or Santorini), the southernmost island of the Cyclades,
with an impressive underwater crater, was also explored at this
time. Many prehistoric remains and ruins from the classical period
were found at Thira, which had been founded before the ninth century B.C.

It would be impossible to mention all the artifacts brought up
from the sea during this expedition. Everything found by the
Calypso team was handed over to the Greek Ministry for Cultural
Affairs and taken in charge by Greek museums. Among the more
interesting objects were a magnificent bronze cannon, pulled from
the French vessel *Thérèse*, found just north of Crete, and many
examples of amphoras, copper plates, marble blocks, and Minoan
cups from over a dozen locations. All in all, thousands of artifacts
were brought up, analyzed, and photographed, to document more
closely the life-style of the classical period in this area.

On September 20, 1976, the *Calypso* team began what would
be their most difficult dives in Greece: dives to explore the sunken
ship *Britannic*. A Galeazzi submersible decompression chamber
filled with a breathing mixture of helium, nitrogen, and oxygen,
large-capacity aqualungs, special cables, slings, and anchors were
only some of the equipment carried aboard *Calypso* for the many
deep dives that would be necessary to investigate this wreck.
Calypso would serve as a diving base during the days but would
shuttle back to the nearest port, St. Nikolo on Kea Island, every
night in order to avoid the possibility of collisions.

Exploration of the *Britannic* was both difficult and costly. The
saucer went down often to explore and film the area; at the same
time, the Galeazzi chamber was submerged to a depth of 130 feet.
Using the chamber as a base, three divers equipped with oversized
tanks filled with the helium mixture descended to the ship for fifteen
minutes at a time. After their assigned chores, they returned to the
chamber to be hoisted up and decompressed aboard ship for a total
of two hours and forty minutes. Due to the extent of these precautions, no diver suffered any ill effect from these very deep dives.

Cousteau's goal in this operation was twofold. First, he wanted
to discover if the *Britannic*, which purported to be a simple hospital
ship, had in fact been illegally transporting war materials, as the
Germans had claimed in 1916. He also sought to find out why this
huge, supposedly unsinkable ship had gone down after only one
blow, and whether this had been from mine or torpedo. To this end,
the *Calypso* team not only relied on primary evidence retrieved
from the wreck but also located and interviewed the remaining survivors from this disaster in order to determine what had actually
happened sixty years before. Unfortunately, the accounts of the accident garnered from the surviving passengers of the ship's last
journey differed drastically; the one thing they agreed upon was that
the *Britannic* was strictly a hospital ship, and was carrying no war
materials of any kind.

One of the biggest prizes of the archaeological expedition was the recovery of this bronze cannon, more recent by far than the Greek remains, but a treasure nonetheless.

Below and right: far more efficient than the earliest version used at Grand Congloué, the new suction hose and fine-mesh recovery basket enable *Calypso*'s team to search large areas and recover even tiny archaeological fragments.

One particularly remarkable survivor agreed to venture out to *Calypso*, and even went down in the saucer to inspect the wreck: Mrs. Macbeth Mitchell, who had served as a nurse on the *Britannic*'s fatal run. This alert and dynamic woman declared that the ship was more beautiful under the sea than it had been in its prime, and jokingly asked the *Calypso* crew to retrieve her sixty-year-old alarm clock from Cabin 15.

Captain Cousteau himself also made two dives with the helium mixture to examine the ship's holds personally. From his and the other divers' reports, he concluded that the *Britannic* was not transporting any war material or military equipment, but was acting as a simple hospital ship when it was sunk by a mine laid a few weeks earlier by a German submarine. Furthermore, they determined that the explosion had hit the ship's coal bunker, which had set off a second explosion of the coal dust. And so the mystery that had shrouded the sinking of the *Britannic* for sixty years was solved at last.

From November 1 to November 23, *Calypso* took the famous archaeologist Lazare Kolonas aboard and returned for the last time to the sunken ship at Andikithira. The suction pipe and air lift were used extensively here to excavate gold jewelry, cut-glass cups, fine pottery and vases, and, most notably, two magnificent bronze statues from the Hellenic Age (the second and third centuries B.C.), all now housed in the National Archaeological Museum in Athens.

It was now time for *Calypso* and her team to return to Monaco, after spending just over a year in the Aegean Sea. Stopping first at the dock in Piraeus, she returned home the way she had come, passing through the Channel of Corinth, the Strait of Messina, and the Strait of Bonifacio. She arrived in Monaco, in miserable weather, on December 1, 1976, and remained there for her traditional winter careening.

A key apparatus in *Calypso*'s study of the Mediterranean was the water sampling bottle, calibrated to collect water at specific depths for return to the surface.

IN THE MEDITERRANEAN

During *Calypso*'s twenty-seven-year history of traversing the Earth, Jacques Cousteau could not have helped but notice the increased level of pollution that was ruining the marine environment he so dearly loved. Ever since his first dive, in 1936, when the waters of the globe had been crystal clear, the fragility of marine ecosystems and the fact that oceans were deteriorating had become more and more obvious to him. Ships had spilled oil into the sea by accident and crews had dumped garbage into the waters on purpose; groupers were virtually eliminated from the Mediterranean; whales had been decimated in the Antarctic; coral gardens in the Red Sea were being choked under tons of waste; and everywhere the most beautiful sites in the marine world were threatened with extinction.

In Captain Cousteau's expert opinion, it had become urgent to establish an impartial and unquestionable "health report" on the state of pollution of the world's waterways. Of course, taking surveys and measurements throughout the waters of the entire world would be an impossible task even for *Calypso*, so, for the purpose of practicality, Cousteau narrowed his objective to examining the conditions of the body of water closest and most familiar to the *Calypso* team: the Mediterranean Sea. According to the plan, results of the survey would be submitted to the United Nations Environmental Program, which would consider the Mediterranean as a scale model of oceans in general; what had happened there had already happened, or would soon happen, everywhere else in the world.

On May 6, 1977, the International Committee for the Scientific Exploration of the Mediterranean, with Prince Rainier III of Monaco presiding, accepted the proposal of its Secretary General, Captain Cousteau, whereby *Calypso* would be devoted to an unprecedented survey of the entire Mediterranean area, including the Black Sea. Although the survey would be conducted under the scientific auspices of the Committee, the Cousteau Society would be responsible for the considerable expenses that such a mission entailed.

The first priority of the survey would be to systematically gather data on the water quality of the sea, to measure exact degrees of pollution, determine the most damaging pollutants, and to choose the most beautiful sites—those deemed most deserving of protection—which could later be turned into national or international parks or biological preserves. But along the way, the *Calypso* team hoped to put their scientific results into more meaningful contexts. How could such a large body of water change so dramatically, become so contaminated, in only forty years? And what needed to be done to restore the oceans to their former vitality? These were just two of the questions they sought to answer as they circled the Mediterranean.

Calypso had been inactive since she had returned from the Aegean in December, but her team had already begun to organize preparations for this expedition. Captain Alinat of the Musée Océanographique of Monaco and Jacques Constans, the Cousteau Society's Vice President for Science and Technology, were made responsible for acquiring all the special equipment required for the taking of samples, recording of information, and measuring of sediment in the sea. Finally, after a total of over six months of research and preparation, *Calypso* sailed from Marseille on July 4, under the direction of Captain Degenne of the Campagnes Océanographiques Françaises. For the first time in history the same ship, with the same crew and standardized methods, was documenting both the beauties of the Mediterranean Sea and the problems that threatened it.

The first inspection took place very close to home; it focused on the area surrounding Marseille itself. *Calypso*—with her divers, troika, helicopter, and diving saucers—operated around the sewers near Calonques and the industrial area of Fos. The side-scan sonar was used successfully to locate the banks of quicksand close to the openings of underwater waste-pipes, and the sounders, compressed air, and the rest of *Calypso*'s equipment were all tested here before the ship moved on to the larger area awaiting her.

On July 27 *Calypso* set out on a long journey, one that would encompass almost eleven thousand miles in the space of less than five months' time. She covered the entire Mediterranean, circling along the coasts and crisscrossing the breadth of the sea. Hour after hour, day after day, *Calypso* cruised, stopped, and took off again. She made 126 different anchorages to take samples of water, sedi-

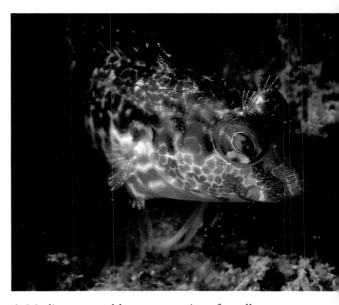

A Mediterranean blenny, a species of small fish that live in warm waters, cruises the bottom for a meal.

A cluster of coral polyps wave their tentacles to feed on floating food; stinging cells line each tentacle to ward off predators.

ments, plankton, and other organisms on the shores of twelve Mediterranean countries. At each stop there was a burst of activity; as the helicopter would take off for an aerial view, Zodiacs would carry team members ashore to study the coastline, and divers would leap overboard to explore under the water's surface. At each site dredges were lowered to collect sediment from the sea's floor, nets were spread to catch drifting animal organisms, and bottles were dropped to various depths to obtain water samples. These samples were then labeled, sealed, and sent off to the marine laboratory for the Atomic Energy Institute in Monaco, where they were analyzed to determine the content of heavy metals and polluting chemical compounds such as PCBs, pesticides, and detergents.

All of the twelve countries *Calypso* visited—France, Spain, Algeria, Tunisia, Italy, Greece, Turkey, Bulgaria, Rumania, Cyprus, Egypt, and Yugoslavia—delegated scientists to participate in the ship's activities, while simultaneously conducting their own research projects. The governments did not always comply with all of Captain Cousteau's requests, however. Often *Calypso*'s travel was limited to areas which were specified by the country's authorities, and to obtain permissions under these restrictions took all Cousteau's patience and diplomatic skill. The work of the crew often had to be done at night as well as during the day, due to time restrictions that were also placed on the expedition. All in all, this proved to be a very difficult expedition for *Calypso*, her crew, and her equipment. At one point, the entire engine system and clutches shut off, and *Calypso* was left drifting in the open sea as a result of stress from having had to tow the troika so slowly on the bottom. Chief Engineer Jean-Marie France had to work day and night to complete temporary repairs that enabled *Calypso* to move on.

Toward the end of the survey *Calypso* paid a second visit to Italy. There, just off the coast of Otranto—an old fishing and resort town—a Yugoslavian freighter, the *Cavtat*, had sunk three years before. She had been carrying a deadly cargo: Six hundred thousand pounds of tetraethyl and tetramethyl lead. For three years, almost five hundred barrels of these lethal chemicals had been resting on the sea floor, largely ignored by the authorities. Now the Italian government was finally beginning to remove the dangerous drums, and *Calypso* went along to check the waters for possible leakage of the deadly poisons.

Calypso finished up the survey by circling around Sicily, then heading home to Monaco, where Captain Cousteau and his team of scientists would help compile and analyze the huge amount of data brought back with them. In general, their results indicated that even if polluting agents had not yet created a real danger in most parts of the Mediterranean, the gradual disappearance of fish and other marine wildlife was a growing and undesirable phenomenon. They identified abusive and unrestricted fishing practices and other forms of interference with nature—such as building artificial banks along

Brilliant color emerges in the dark depths of the Mediterranean under the lights of a *Calypso* camera team.

the shores, altering the courses of rivers, and overdeveloping human uses of the coasts and waters—as the major factors in the decline of marine life in the Mediterranean Sea.

Although the results of the mission did not produce any easy solutions to the problems of pollution, they did indicate that a greater balance between the commercial and natural worlds was needed. The mission also alerted the leaders of many different countries to the problems that faced their shores and waterways, and proved the need for continued study of the sea as well as the increased responsibility for its condition by the nations that bordered it.

ON THE RIVER NILE

As the year 1978 began, Captain Cousteau had no firm plans for *Calypso* except that she was scheduled to receive a series of major repairs in the fall. Even while the ship remained idle, however, the *Calypso* team did not—they used the time to further explore the polluting agents that were affecting the Mediterranean Sea.

Their year-long in-depth study of water conditions in the Mediterranean prompted them to undertake a similar investigation of the Nile River. As the major source of water to the Mediterranean Sea, the Nile imparts to it indispensable life-sustaining nutrients. The Nile's condition is crucial to that of the entire Mediterranean area. The next logical step following the study of the previous year, then, was to find the exact extent of the relationship between river and sea; additionally, the *Calypso* team sought to discover the rela-

Calypso II, the PBY Catalina flying boat, served as the principal vehicle for the Nile river expeditions.

tionship between this all-important river and the peoples who depended upon it for their existence.

Philippe Cousteau and fourteen other members of *Calypso*'s team set out in February 1978 on a ten-month expedition covering the full four-thousand-mile length of the Nile. Although *Calypso* herself remained docked in Monaco, much of the equipment associated with her went along on this expedition, including several four-wheel-drive land vehicles, inflatable boats, and the *Flying Calypso*, her new twin-engine PBY Catalina seaplane. (The *Flying Calypso* had a 3,800-mile range and a maximum altitude of almost 20,000 feet, and would prove itself invaluable on this and future missions.) Sample collections and pollutant measurements were made at every significant location: tributaries of the river, outlets of the main stream, and spots above and below every important commercial or agricultural center.

The findings of this extraordinary scientific survey were not encouraging. In fact, the *Calypso* team's investigations revealed that in effect the Nile no longer even flowed into the Mediterranean Sea, no longer provided that body of water with its life-giving properties. Instead, the Nile River was now being used up entirely in hydroelectrical or irrigation projects, its nutrients deposited at the bottom of dam-created artificial lakes instead of on the Mediterranean's sandy floor.

Meanwhile, Captain Cousteau had decided that it would not be prudent to let *Calypso* stand in dock for a whole year, so he improvised a program to occupy the interim between the spring and fall repairs. He made plans to send the ship out on a search for wrecks in the Marseille area. This way, her team would be able to photograph at a leisurely pace in order to complete a new film on shipwrecks that the Captain had been wanting to make for a long time. One of his first films, in 1946, was about shipwrecks—it was this film that had first captured the public's attention—so the subject held a special meaning for Cousteau.

In the lore of the sea the word "shipwreck" has magical meaning, full of drama and mystery, evocative of an exotic and disturbing past. Jacques Cousteau considered sunken ships as witnesses to the past, witnesses that could bring the past to life.

Cousteau focused his interest on finding a shipwreck believed to hold a fantastic treasure, as the romance associated with it would then be even greater. One such ship, the *Natal*, had been brought to Albert Falco's attention many years before. Launched from France in 1881, this ocean liner was sunk on the night of August 30, 1917, following a collision with a cargo ship. The tragic accident took more than one hundred victims.

Much was known about the circumstances of that fatal night, but tales full of mystery and intrigue were still being told. It was rumored that the *Natal* had been carrying gold ingots—fifty-five cases of them. Given the price of gold in 1978, the treasure buried

Clouded by sediments, Nile water proved impossible for underwater photography, so fish trapped for study were filmed and observed in a portable, clear-walled tank. The fish shown here is one of several species that form a basic diet for the Shilluk people who live beside the Nile and net their prey.

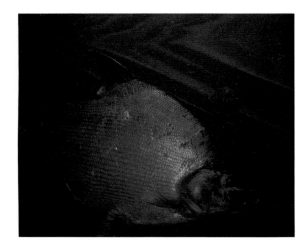

Calypso's land team surveyed all forms of wildlife in the river valley of the Nile.

Overleaf: along with biological treasures, the Temple of Ramses II at Abu Simbel was threatened by the construction of the Aswân High Dam, which created a huge lake in Egypt and the Sudan. Though lesser sites were flooded, Abu Simbel was rescued through a gigantic international effort.

with the *Natal* would be valued at over $15 million. A few fishermen had tried to salvage the cases, but their boat had overturned in the attempt and everything was lost. It was said that other fishermen had found some gold coins in trawling nets, but this was never proved. Cousteau knew well that in dealing with buried treasure one can never be sure what, if anything, is fact, and what is mere romantic dreaming.

In order to find out more about the *Natal* it would be necessary to investigate the wreck firsthand. But first, it would be necessary to locate it. And before that could be done, the *Calypso* team needed to learn exactly what the *Natal* had looked like. In the beginning, they had only a drawing of the ship found on an old postage stamp. Further, they had only a very general idea where the wreck had taken place. As time went on, however, Alexis Sivirine managed to locate the ship's blueprints and some photographs, and Albert Falco, by talking to fishermen in the area, discovered the sites of some sunken ships near where the *Natal* was said to have gone down.

Calypso left Monaco on July 9, after many months had been spent researching the *Natal* on land. In addition to her usual equipment—side-scan sonar, helicopter, and diving saucer—*Calypso* was carrying one of Dr. Edgerton's newest inventions: an EEG sonar. This instrument goes down to depths of approximately 300 feet and sends up a picture of the bottom of the sea. It was designed to scan an area up to 900 feet across.

During the weeks that followed, *Calypso* cruised continually around the Marseille area, following a course that was determined by the range of the radar and sonar equipment. The team discovered several small sunken ships along the way, none of which turned out to be the *Natal*. The first wreck they came across was identified as a small coastal freighter probably sunk in World War I. Another ship sunk in that war, an auxiliary cruiser known as the *Drome*, was located near the first. Then two more wrecks, one dubbed the *False Natal* by *Calypso*'s team and one unusual paddle-wheel boat called the *Logobo*, were also inspected during July.

The *Calypso* team were beginning to despair of finding the *Natal* among so many wrecks, when finally, on August 2, a picture appeared on the EEG recorder that seemed to be of the ship they sought. Captain Cousteau and Albert Falco went down in the saucer to the site of the wreck and came back to the surface with the exciting news that it was indeed the *Natal*. Unfortunately, the wreck lay more than 300 feet below the surface and was thus too deep to be visited by scuba divers. Descents in the saucer allowed inspection of the outside of the hull, but no one could venture into the ship's interior. The wreck was lying on a bed of sand; the upper part of the ship had exploded, and sheet metal now littered the sea floor. During subsequent dives, the saucer circled the wreck several times and found the wood on the deck decayed, the masts and smokestack knocked down; a gaping hole in the hull allowed the divers to glimpse the first-class cabins, where suitcases, still packed, were seen. There was no sign of the gold ingots they had hoped to retrieve from the wreck, however, or of any other object of value. A few pieces of the wreckage were gathered up by the saucer's "arm" and brought on board, but then it was time for *Calypso* to return to her home port to undergo her scheduled repairs. The treasure, if any, remained below.

Calypso arrived in Toulon on October 23, when a major careening began in which the ship was completely re-waterproofed, caulked, painted, and inspected. On December 8, Alain Vichon took over as captain, and took *Calypso* out for a few days of testing before returning to the shipyards, where work on the ship continued until March 1979.

W hile *Calypso* was under-going a complete reconditioning, including the replacement of her entire electrical system, the redesigning of her chart room, and modification to her many existing telecommunication systems, a plan was brewing that would change the course of her future.

For some years Captain Cousteau had been growing weary of the constant struggle he faced in procuring funding for each *Calypso* expedition, preferring to concentrate his energies in areas where he was more effective. Because of the great success being enjoyed by the Cousteau Society in the United States and support offered by the leaders of the city of Norfolk, Virginia, where the Society's headquarters were located, Captain Cousteau now considered moving his offices to America. His activities would be easier and less costly there than they were in France. In 1979 he decided once and for all to transfer his base of operations from Europe to Norfolk, keeping liaison offices in France and Monaco.

Therefore, on April 21, *Calypso* left Monaco and sailed, under the direction of a new captain, Alain Traounouil, to what would be her new home in the United States. The crossing was largely routine, notable only for the persistent fog, and was broken by stops at Gibraltar (to pick up her helicopter) and the Canary Islands (to replenish food supplies).

Calypso arrived in Norfolk on June 2, 1979. She was given a tumultuous welcome there, which included a military band playing the *Calypso* hymn—a song written and popularized by John Denver, a member of the Cousteau Society's Council of Advisors. As soon as the festivities were over, the team turned to the monumental task of unloading the many tons of equipment *Calypso* had carried across the ocean to her new home port.

Eight days later *Calypso* commenced her first American mission: to explore and photograph the wreck of the United States ship *Monitor*. The first ironclad, revolving-turret battleship in the world, the *Monitor*, built in 1862, made battleship history when it fought against the C.S. *Virginia*, formerly the *Merrimack*, during the Civil War. Unfortunately, the *Monitor* sank the same year it was built— "swallowed by the sea from the rear," as the accounts of the day reported—while it was being towed back northward. The tragedy

A New Home for Calypso

In order to locate the wreck of the *Roraima*, *Calypso* team members lower a side-scan sonar towfish into the waters off the coast of Martinique.

Top: Haemulon parrai, popularly known as "grunts" due to the sounds they make by grinding their teeth together.

Opposite: Calypso leaves her new base port at Norfolk amid a rain of balloons.

occurred just off Cape Hatteras, at night, and claimed the lives of sixteen men.

Since Cape Hatteras is a mere 150 miles from Norfolk, and at the request of the local authorities, Captain Cousteau decided to send his crew to visit the wreck. Although the water was not clear, the wreck of the *Monitor* was spotted quickly and easily, over 200 feet below the surface. Eight divers, including Philippe Cousteau, went down several times and, despite a strong current from the Gulf Stream and other difficulties inherent to all work at sea, managed to examine and film this historic sunken ship.

Her mission accomplished, *Calypso* left Cape Hatteras and headed toward Martinique in the Caribbean. Her voyage was unexpectedly interrupted on the morning of June 19, when the *Calypso* team were informed of a shocking and tragic event: Philippe Cousteau, thirty-nine years old, the Captain's younger son, had died in an accident during a routine testing of the *Flying Calypso* seaplane in Portugal.

Although the grief remained, the shock of Philippe's untimely death abated somewhat in the next days, and life aboard *Calypso* went on. She continued on her way to Martinique, where the eventual destination was the town of St. Pierre. St. Pierre is situated at the foot of Mont Pelée, the site of a disastrous volcanic eruption in 1902. Ships that were anchored in the bay at St. Pierre when the disaster struck had sunk immediately to the bottom like lead weights. An opportunity to film so many sunken relics of the past could not be passed up by the *Calypso* team. Cinematographer Henri Alliet

Among the Venezuela experiments were several in human physiology. Here, the diving team monitors the condition of a subject in the decompression chamber.

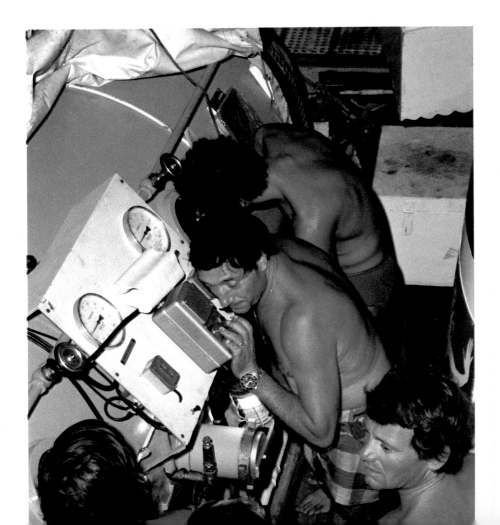

166

joined Albert Falco and his divers, and in scuba gear as well as from the diving saucer they filmed the *Rorema*, the *Gabrielle*, the *Teresa de Vigo*, and other living witnesses of the volcano's tremendous force.

On July 25 *Calypso* set out on her next mission, four hundred miles to the south. This mission was similar to those conducted in the Mediterranean in 1977 and the Nile in 1978 in that it too concerned the rigorous control of water pollution, but now the *Calypso* team were concentrating their activities off the Venezuelan coast. Their study would extend from the Gulf of Venezuela to the mouth of the Orinoco River.

On July 27 the ship was in Cumaná, Venezuela, the oldest permanent European settlement in South America, ready to begin a three-month intensive study of the area's waters. Jacques Constans and Albert Falco were codirectors of the mission, for which they were joined by scientists from seventeen different Venezuelan institutions delegated by the government. The area studied was of critical importance to all Caribbean countries, for it is just off Venezuela that turbid but rich waters from the Amazon and Orinoco rivers meet with the nutrient-poor waters of the Atlantic Ocean. The mixture flows between Grenada and the Venezuelan coast and affects the fishing activities, biological productivity, and water clarity of ten nations.

Calypso's mission in this area was divided into two phases. The first was devoted primarily to hydrology, geology, and the study of life forms such as plankton. The ship set sail from Cumaná and headed toward the mouth of the Orinoco River, then on to a point in the Atlantic close to Barbados, then back to Puerto La Cruz on the mainland. From there she journeyed to the Gulf of Venezuela and finally to Maracaibo.

During this first phase of the investigation *Calypso* logged 588 hours at sea, of which 168 were spent at 110 different scientific stations. She voyaged a total distance of 3,130 nautical miles and took hundreds of water sediment and plankton samples, as well as three-foot-long geological cores. The data obtained were later turned over to the Venezuelan scientific community for detailed study.

The second phase of *Calypso*'s Venezuelan mission began in Maracaibo at the end of August and lasted until the end of October. Work during this period was concerned mainly with geological and biological studies, most notably of the area's fisheries and coral reefs, and provided the sciences with much new information about the Caribbean's marine life.

Perhaps the greatest moment of the trip took place on September 24 at the Blanquilla Islands north of Margarita Island. It was here that the SP-350 diving saucer reached a total of one thousand successfully performed dives. Piloting the saucer on this momentous occasion was Raymond Coll, a fitting choice since he had served with Captain Cousteau for over twenty years. Celebration of the event included champagne for everyone, from bottles that had been stored in *Calypso*'s hold for several years for just this occasion.

Dwarfed by the sheer granite cliffs of the Saguenay River fjord, *Calypso* makes her way from the St. Lawrence. Highlight of the exploration was a 600-foot dive in the SP-350 made by Jean-Michel Cousteau and Albert Falco, who discovered, after descending about 50 feet through water that ran the gamut of shades of chocolate, a thick layer of clear, cold salt water that hosted numerous forms of brilliantly colored life.

All in all, the journey to Venezuelan waters was a very successful one for the *Calypso* team. Approximately 400 work stations, 222 sounding and dredging operations, 25 dives in the diving saucer, and 38 flights in the helicopter resulted in more than 1,350 samples and tests. From this data a preliminary report was formulated—recommendations on how to clean up the area and which specific sites needed the most immediate attention—and presented to Venezuelan authorities.

The mission successfully completed, *Calypso* sailed on November 1 for Curaçao in the Lesser Antilles, where she was scheduled for her regular careening. Here, in addition to routine maintenance work, she received a new device designed to facilitate difficult maneuverings in tight spaces: an electrical propeller attached to her false nose. On November 11 the ship set out again, this time for Fort de France in Martinique, arriving there two days later. From Martinique most of the members of the team flew home to enjoy a rare extended break.

Four months into 1980 *Calypso* was still in Fort de France, Martinique. This was the most extended rest period enjoyed by the ship or the members of her team, most of whom had gone back to France for a well-deserved leave. Such was not the case for Jo Seguy, however, who used the time to wash, scrape, and paint *Calypso* from her stem to her rear crane. Sparkling clean, painted black-

and-white, and without a speck of rust, *Calypso* was once again ready to face new missions.

By the beginning of April the crew members had all returned to the ship in Martinique and began to ready her for the trip north to Norfolk. She was still under the command of Captain Traounouil and sailed with several new crew members, among them an American officer called Captain John, whose task was to keep the watch until the ship reached Norfolk. Jacques and Simone Cousteau were also aboard for the return trip.

Calypso started northward on April 15, taking advantage of an unusually strong Gulf Stream current that seemed to give her wings. She traveled 2,130 miles in only nine days, without suffering any mechanical or weather problems. This was indeed a remarkable performance for her, made even more remarkable by the fact that she had to make a detour to stop in Miami in order to drop off Captain Cousteau so that he could fly to New York.

Once in Norfolk, *Calypso* looked and felt right at home berthed alongside a dock near the city. The public exhibited a great deal of interest in the little ship, now the most celebrated research vessel in the world, and never seemed to tire of looking at her. In fact, she was one of the star attractions of the Norfolk Harborfest in the beginning of June. Here she was paraded among thousands of ships, as divers went down in scuba gear and the diving saucer to demonstrate their underwater equipment and entertain the crowds. The entire staff of the Norfolk Cousteau Society, with some assistance from the Society's New York City staff, spent four days demonstrating *Calypso*'s activities and leading tours of the ship for visitors during the Harborfest.

AN INLAND PASSAGE

Calypso was still awaiting a new mission. A project slated to take place in the Amazon River area was in the works, but not yet near fruition. Captain Cousteau decided to carry out another program in the meantime: an ecological survey combined with the shooting of three films to be coproduced by the National Office of Canadian Film. The settings for these films would be unlike any ever visited by the *Calypso* team: the Greater Saint Lawrence Waterway, including Newfoundland and the Canadian Great Lakes.

Preparations for the expedition took two months. Advance work performed by three Society staff members, Karen Brazeau, Paula DiPerna, and Laurie Wolfe, produced a list of Canadian scientists and institutions eager to contribute to *Calypso*'s work along the way. The same three women also went off on research trips throughout Canada to select and identify the sites of greatest scientific interest in the region, as well as to work out the logistics of the expedition. Aboard *Calypso*, expedition work was being directed by Captain Cousteau, Jean-Michel Cousteau, Jacques Constans, and expedition leader Albert Falco. Dominique Sumian was put in charge of all specific explorations once *Calypso* reached Canada,

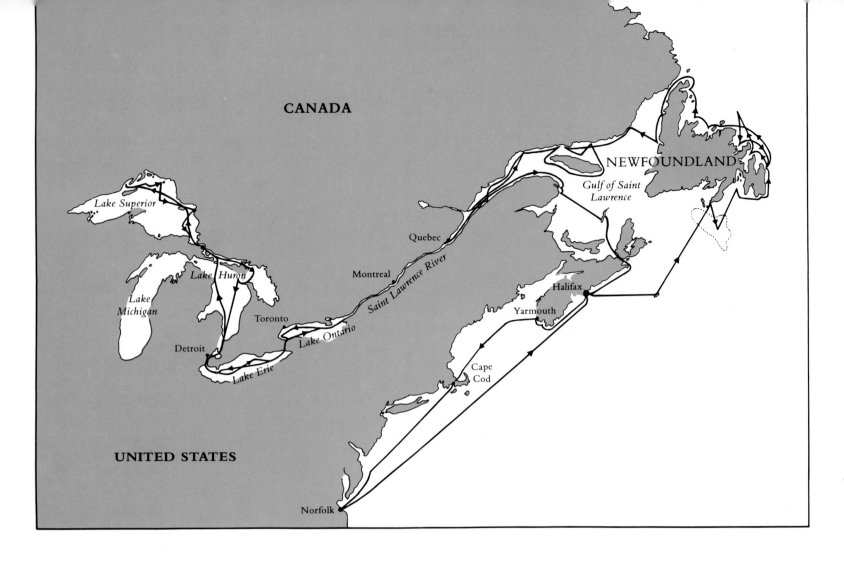

CANADA

NEWFOUNDLAND

Gulf of Saint Lawrence

Lake Superior

Lake Huron

Lake Michigan

Quebec

Montreal

Saint Lawrence River

Halifax

Toronto

Lake Ontario

Yarmouth

Detroit

Lake Erie

Cape Cod

UNITED STATES

Norfolk

In Labrador, chief expedition diver Dominique Sumian befriends a baby fur seal.

Opposite: near Bay Como, on a tributary of the St. Lawrence, the film crew approach dangerously near a thundering waterfall.

170

and, as usual, Simone Cousteau was on board for the entire voyage, seeing to the many needs of the crew.

On June 28 *Calypso* arrived in Halifax, Nova Scotia, where she was hospitably welcomed by the Bedford Institute of Oceanography. While at Halifax, Marc Soviche took over the charge of the ship from Captain Traounouil.

The team left Halifax on July 2 to begin investigations and filming on land—on Sable Island, ninety-five miles east of Nova Scotia—and at sea. Here they went underwater to explore a fabled graveyard of sunken ships, then, back on the crescent-shaped island, surveyed the only nesting site of the far-ranging Ipswich sparrow. Some whales of the area felt comfortable enough with *Calypso* to approach and inspect the ship at anchor.

The next stops on *Calypso*'s agenda were St. Pierre and Miquelon, two tiny French possessions that are the only remnants of all the territories France once owned in this area. After visiting with their compatriots on the islands, the diving team went undersea to examine the shipwrecks nearby. The team headed toward nearby Fortune, in the south of Newfoundland, taking along a sailor from St. Pierre, Bruno Vidal, who would later become *Calypso*'s first mate. Then they returned to St. Pierre to film the activities of a fishing ship there, the *Croix de Lorraine*.

Calypso continued her mission by sailing around Newfoundland counterclockwise. Her next stop was at Placentia Bay, then it was on to Trepassey Bay, where the diving team studied a recent but increasingly menacing problem: the entanglement of humpback whales, an endangered species, in cod-fishing nets.

Study of this problem was continued at St. John's, where *Calypso* put in to have repairs made on the hydraulic crane, essential to launching and retrieving the diving saucer. While docked here, the team heard of a humpback that had just been caught in a fisherman's net. The divers loaded a Zodiac, their diving gear, and much bulky camera equipment into a rented truck and sped to the site. They found a young humpback struggling, the main rope of the net stretched across its mouth like a bridle bit. The whale had obviously put up quite a struggle against the net, for its delicate skin was scratched and bleeding. The creature seemed to have given up all hope of freeing itself by the time the *Calypso* divers arrived, and concentrated its efforts simply on returning periodically to the surface for long gasps of air.

Calypso's Bernard Delemotte immediately set his mind on rescuing the whale. The first step toward achieving this was to gain the animal's acceptance, so he and the other divers went down and for an hour and a half attempted to reassure it. By drawing his palm slowly across the whale's lips he tried to convince it to open its vast mouth to free the rope, but to no avail. Then he pulled out his knife

Leaning on the "roof" of a beaver dam, Jean-Michel Cousteau hears a report from divers who have been investigating the underwater portion of the structure at Michipicoten on the eastern shore of Lake Superior.

and cut the rope, pulling it with all his strength by setting his feet on the whale's side and arching his back. He finally managed to extract the rope and liberate the frightened young whale. In what appeared to be a show of gratitude to Delemotte, the whale let him climb onto its back, circled around the area several times, then headed slowly out to the open sea. Delemotte rode the whale for about a mile. All this was captured on film by divers and team members on the Zodiac.

From St. John's, *Calypso* continued on her way around Newfoundland, stopping at Trinity Bay on July 25. At tiny Funk Island, the team filmed a group of scientists conducting a census of seabirds—murees, puffins, gannets, and kittiwakes. Stops were also made at Cape Bonavista, the Newman Sound, and at Red Bay in Labrador. The *Calypso* team explored both the land and the sea, and the diving saucer made an additional twenty dives.

On Newfoundland's western coast *Calypso* laid over in Riche Point and then Bonne Bay before proceeding to the mouth of the St. Lawrence River, in many ways the most important waterway in North America. Havre St. Pierre was reached on August 6, and *Calypso* filmed the natural surroundings there as well as several shipwrecks, using the diving saucer, Zodiacs, helicopter, and the new hovercraft recently brought aboard. The ship continued west, sailing farther up the St. Lawrence. On August 16, at midnight and under the northern lights, the saucer went down to a depth of almost 650 feet. The rest of the crew, on the surface, spent thirty minutes watching the celestial display while they waited for the saucer to resurface. As they waited, they heard a loud commotion in the water, apparently caused by dolphins. The sounds were so intense that they prohibited telephone communication with the men in the diving saucer. The cause of this impromptu concert was never determined; perhaps it was the northern lights, or maybe a prodigious amount of shrimp at the bottom, the presence of which was confirmed by the divers in the saucer.

Then *Calypso* was off again, wending her way farther up the river, with ports of call at Pointe-des-Monts, Comeau Bay, St. Pancrace Bay, and at length Rimouski. There, along the southern banks of the St. Lawrence, the team went down in the diving saucer to explore and film the wreck of the *Empress of Ireland*, a luxury liner that sank in 1914 after a collision with a collier, causing the loss of more than a thousand lives.

Next, it was on to Bic Island and to Tadoussac, located where the Saguenay River joins the St. Lawrence. There, between impressive cliffs, an extraordinary upwelling of nutrients in the water attracts shoals of fish and many species of whales. The crew filmed hundreds of beluga whales, five blue whales, and untold numbers of fin and sei whales.

Continuing her journey, *Calypso* passed Quebec and arrived in Montreal on August 29. After a two-day layover the ship proceeded

More accustomed to flexible diving flippers than rigid snowshoes, Jean-Michel Cousteau, Bernard Delemotte, and Dominique Sumian prepare to set out overland for the shore of Lake Superior.

through the floodgates that opened into Lake Ontario. She anchored at Kingston in eastern Lake Ontario on September 3 for a quick and routine exploration of a sunken three-masted schooner, the *Barnes*. During what seemed like a particularly easy dive, tragedy struck the expedition in the form of a diving accident. At the end of a thirty-minute session underwater, one of the divers, electrician Rémy Galliano, a member of the *Calypso* team for over a year, failed to surface. A search party was organized immediately, but they could find no trace of Galliano. Finally, five hours later, with the help of divers from the Canadian Armed Forces, Galliano's body was discovered on the muddy lake bottom.

Diver Raymond Coll makes ready to slip below the ice for a study being conducted of local commercial fishing practices in winter at Sainte Anne-de-la-Perade in the St. Lawrence west of Quebec.

His death was later attributed to air embolism, a condition that occurs when air expands in a diver's lungs, rupturing the lung tissue and forcing air into the bloodstream, thereby resulting in a reduction of oxygen and profound damage to the brain. It was suggested that perhaps Galliano had misjudged his depth in the muddy waters of Lake Ontario and so had ascended without exhaling, a fatal mistake which had led to an air embolism. Although this condition is a recognized hazard of diving, the *Calypso* team since 1952 had logged more than thirty thousand dives without an accident.

The accident deeply affected the entire crew, all of whom regarded Rémy Galliano as a close friend and part of the *Calypso* family. He was just thirty years old when he died, a cheerful, genial young man, and his loss was deeply felt by his fellow divers.

Heavyhearted, the team nevertheless resumed their activities the following day. On September 6, the ship was in Newcastle; the next day she dropped anchor just off Port Weller, where the team dove down to visit the wreck of the *Scourge*, a perfectly preserved commercial schooner commissioned during the War of 1812 and loaded with cannons and other weaponry. It had earlier been discovered by an American dentist in icy waters at a depth of over 280 feet. Unfortunately, the *Calypso* team had to terminate the dive due to lack of visibility; although the wreck was detected by the sounder, it could not be seen from the saucer even from a distance of six feet.

On Wednesday, September 10, *Calypso* passed a number of floodgates and entered into Lake Erie, which she crossed in one day. She then went down the Detroit River, which took her past Detroit, to Lake St. Clair and finally emerged into Lake Huron. On this part of the trip, *Calypso* was accompanied by a flotilla of four hundred boats, small and large, that came to greet and escort her. After crossing Lake Huron, she arrived at Canada's Sault Sainte Marie, which gave access to Lake Superior.

At Thunder Bay, on western Lake Superior, the farthest point reached in the expedition, the team filmed a key exchange point

Parka-clad scientists prepare to trail a plankton net in the Saguenay River.

Above and opposite: all hands turned to for ice-clearing after *Calypso* encountered a ferocious storm in the seventeen-mile-long Strait of Canso between Cape Breton Island and Nova Scotia.

Overleaf: the most northern of the Canadian explorations of the *Calypso* team took place on land by snowmobile at James Bay, the southernmost extremity of Hudson's Bay. Here, Jean-Michel Cousteau and Bernard Delemotte ride beside the overflow canal of a hydroelectric dam.

along the early fur traders' routes. Evidence of Indian pictographs now submerged, and of a sunken fur barge, was documented by the crew. In deep water just outside the bay, the diving team found to their astonishment a habitat nearly devoid of life, although many species of fish are common to the lake. The absence of flora and fauna in this area suggested that the vitality of Lake Superior was suffering, perhaps coming to an end.

The return trip began on September 16. *Calypso* took the longest possible route home, however, sailing around islands known for their natural parks and stopping often at points of interest. Near Copper Island in Lake Superior, the team filmed a yacht that had sunk in 1911, the *Gunilda*. The boat was extraordinarily beautiful, with its brass and mahogany trim and its gold-painted bow still intact.

At the Slate Islands, in the far north of Lake Superior, the team filmed and studied the behavior of caribou, about three hundred in number, which swim from island to island to find food and mates. These elegant animals are now in danger of extinction owing to a

variety of causes, especially the curtailment of their habitat by human encroachment.

In southeastern Lake Superior *Calypso* documented the wreck of the *Edmund Fitzgerald*, an enormous steamer that had sunk so quickly in 1975 that its crew did not even have time to send a distress signal. The 729-foot ship, visited by the diving saucer, lay at a depth of 520 feet and was split into two pieces that were found 150 feet apart.

On September 24 *Calypso* reached Sault Sainte Marie again, this time stopping to visit the Sea Lamprey Control Center, part of the Canadian Department of Fisheries and Oceans, where they documented the massive control program designed to rid the Great Lakes of this "invader species." (The sea lamprey is a parasite that attacks and lives off valuable species of food fish.) Then they were off again to Detroit, where Captain and Jean-Michel Cousteau were met and interviewed by city officials and journalists from all over. That same evening, September 30, *Calypso* left Detroit, crossed Lake Erie, sailed through the Welland Canal, and arrived at Port Weller. There she returned to the site of the *Scourge* for three more dives in the saucer. This time the water was much clearer, allowing the camera crew to successfully document the excellent condition of the ship.

The end of the expedition was approaching, and on October 6, the one hundred and sixth day of the Canadian mission, *Calypso* arrived in Montreal, where a maintenance crew would spend the next six weeks working to overhaul the ship in preparation for a winter sail down the ice-filled St. Lawrence. Meanwhile, most of the team members, who had worked the last three and a half months without a single day off, flew home to rejoin their families for a few weeks' leave.

The overhauling work complete, on Friday, November 28, *Calypso* departed Montreal under the supervision of Captain Traounouil. The weather was cloudy, the visibility was poor, and snow was falling as she made her way back down the St. Lawrence River. Several times during the return trip the ship had to put in unexpectedly at ports such as Quebec City and, much farther downriver, Sept Iles, so that thick accumulations of ice could be scraped from her decks, equipment, and lines.

Finally, on December 23 *Calypso* pulled into her home berth in Norfolk, Virginia, and the Canadian expedition officially came to a close. She had weathered the treacherous conditions, but had sustained severe damage to her hull and engines while negotiating the narrow and icy St. Lawrence and its gulf, both with extremely strong currents. At one point, fifteen tons of ice had clung to the little research vessel, throwing off her balance and threatening to sink her. But now she was safely home for Christmas. Preparations were already being made for next year's journey, *Calypso*'s most exciting expedition ever: a trip to the Amazon.

T he plan, a dramatic one, had evolved in Captain Cousteau's thoughts over many years. It was this: to send *Calypso* on a year-long mission to the Amazon, the world's largest river, and thus into the world's least-known and most forbidding jungle.

The opportunities for scientific and cinematic exploration in this continent-sized river basin would be overwhelming. Nearly *one-fifth of all the river water in the world* flows in the Amazon, making it the most important freshwater root system of the sea. More than a thousand tributaries, ten of them larger than the Mississippi, thread through the dense rain forests of eight countries to join in a single brown flood so colossal that its discharge penetrates two hundred miles into the Atlantic. The river and forest comprise what is undoubtedly the richest and most diverse biological treasury on earth. The bewildering assemblage of exotic plant and animal life dwelling in the Amazon jungle, earth's largest rain forest, is matched by the unparalleled array of aquatic life darting through the murky waters that drain some 2.7 million square miles between the Andes and the sea. There are more species of fish in the Amazon than in the Atlantic Ocean. And there are freshwater cousins of some of the ocean's most intriguing animals, including dolphins, sharks, turtles, sting rays, manatees, and turtles. The world's largest otters live in the Amazon, and the world's largest freshwater fish; and catfish so large that they have been known to eat children.

By Captain Cousteau's estimate, a journey into Amazonia would amount to *Calypso*'s most complicated and most ambitious program ever. Unfortunately, the scope of the mission would also make it the team's most expensive venture. Not only would the demands of logistics, personnel, and equipment be enormous, but the ship—badly damaged by ice storms during the Canadian expedition—would need extensive repairs and renovation.

A program of rehabilitation was devised for the ship while funds were sought to prepare for an Amazon voyage. Chief Engineer Jean-Marie France directed the work of overhauling the ship's main engines and mechanical systems; Chief Diver Raymond Coll supervised the exhausting work of scraping, sanding, repairing, and repainting the entire ship; and engineer Gaston Roux began to repair and rebuild three of *Calypso*'s diving saucers. The work proceeded throughout the spring and summer of 1981, most of it

Calypso Countdown: Rigging for the Amazon

Top: Calypso, her crew, and equipment have come a long way since this diver was photographed in the Red Sea in 1963.

Opposite: with the vast flood plain of the Amazon stretching into the distance, *Calypso* sails upriver to a major new expedition.

At dry dock in Norfolk *Calypso* receives a complete overhauling to prepare for her long Amazon expedition; special preparations are needed for work in tropical water and on-board conditions of high heat and humidity.

carried out by members of the crew in order to save precious funds.

Then, in mid-1981, the financial backing for an Amazon mission was proffered by the entrepreneur and cable-TV magnate Ted Turner, who was introduced to Captain Cousteau by entertainer John Denver. (Denver had long been a conservationist and supporter of Cousteau Society projects.) Turner and Cousteau reached agreement on a contract calling for the production of at least four hours of television documentaries covering the Amazon, and the dream suddenly became a reality.

Now the scope of the renovation work on *Calypso* expanded. Jacques and Jean-Michel Cousteau decided that a major overhaul of the vessel—the most extensive rehabilitation in fifteen years—would be carried out to prepare *Calypso* not for one but for five years of work. Their long-range plan was first to explore the Amazon, then set out for the South Pacific, and, eventually, to investigate China. The aging wooden ship would need a thorough renovation before departing even on the first stage of her journey.

Before departure from Norfolk, every piece of diving, ship's, and scientific equipment was checked over. Here chief scientific officer Jacques Constans *(left)* and Albert Falco work on a water-sampling probe.

Because *Calypso*'s flanks had been deeply scarred by ice floating in the St. Lawrence, the ship entered a Norfolk dry dock where workmen inspected and repaired her hull. Gouged hull planks were removed along both sides of the bow and replaced by some 450 linear feet of heavy planking. The remainder of the hull was stripped clean by blasts of an abrasive stream of ground walnut shells, then patched, caulked, and coated with an antifouling paint. Port and starboard propellers were removed, cleaned, and reconditioned to their original pitch and diameter, then balanced and reinstalled. Both propeller shafts were also cleaned and overhauled. (Unbeknownst to the ship's crew, the two shafts would ultimately break, causing tremendous problems when *Calypso* finally set out for South America.)

Another major project centered around the ship's main and auxiliary engines. Under Jean-Marie France's direction, the two General Motors straight-eight-cylinder diesel main engines—veterans of nearly four decades of work—were completely disassembled and rebuilt by a team of engineers that included Philippe Lagadec, Paul Martin, John Masiak, and Joe Cramer. From her office in New York, *Calypso*'s logistics coordinator, Susan Spencer-Richards, was able to obtain most of the spare engine parts, as well as scores of other ship's parts, as gifts to the Cousteau Society. Among the donations arranged by Spencer-Richards were two new auxiliary engines, which arrived in October. The "6.71" six-cylinder motors were presented to *Calypso* as a gift from the Detroit Diesel Allison corporation. Affixed to one of them was a small plaque from the workers who assembled the motors, expressing admiration for the ship and its campaigns. Nearly two hundred employees of Detroit Diesel—everyone who had helped in the construction of the two motors—also signed a letter of goodwill and good luck to Captain Cousteau and the crew.

The list of projects the crew must complete before *Calypso* could set out for Amazonia was formidable: major rebuilding, not only of the engines, but of the transmissions, the water-coolant systems, the water tanks (including installation of two new reverse osmosis, one-thousand-gallon-a-day "watermakers"), the exhaust system, the compressors (one of which was donated by Ingersoll-Rand), the bow thruster, and the electrical motors and generators (some thirty-two in all, which power a variety of pumps, hydraulic systems, winches, and steering mechanisms). Because melting ice had descended through the decks of the vessel during the Canadian expedition, raining down on the main electric panels in the engine room and attacking relays and circuit breakers, the panels had to be completely cleaned and rebuilt.

As this heavy work was carried out, the crew also rehabilitated and improved other systems aboard ship. Guy Jouas and Tom Dettweiler thoroughly renovated the electronics equipment, including satellite navigation and communications systems. Raymond Coll and fellow divers Jacques Delcoutère and Marc Zonza repaired diving equipment and added improvements, including a high-pressure breathing–air purification system donated by Bauer Breathing Air. Two new freezers provided by Amana were installed below deck, one for the forty-five-day food supply, the other for storage of scientific samples. In anticipation of a year in the hot and humid Amazon, air conditioners and an ice maker were added to the ship, as well as new canvas awnings and equipment covers.

The unique requirements of the Amazon mission, involving travel by land and air teams dispatched from *Calypso* across vast distances, as well as the challenges of a river that originates in icy Andean mountain streams and terminates in steamy jungle, compelled the crew to add special new equipment. During the fall and winter of 1981 *Calypso*'s complement of sophisticated tools was increased by a six-passenger hovercraft, a thirty-seven-foot inflated rubber raft (for travel in the rapids of the Peruvian Andes), and a six-passenger Cessna "bush plane" equipped with floats. In Italy, Cap-

At the mouth of the Amazon, in Belém, Brazil, an important piece of supercargo is hoisted aboard: the dragon-headed wooden sculpture is a *carranca*, a totem believed to keep evil spirits away from watercraft.

185

tain Cousteau and Jean-Michel inspected an amphibious IVECO truck, only the second model ever built, and arranged for its participation in the Brazilian Amazon missions—a program financed by IVECO.

Arrangements were also made with the IVECO Truck Company to ship a special six-wheel-drive truck—like a model that had proved valuable in Canada—to Peru, where a land team was to use it in exploring the Andean headwaters of the great river. The huge yellow vehicle, quickly named "Amarillo" (Spanish for "yellow"), was extensively modified for the jungle and high-mountain expeditions. A powerful crane was fastened to the side of the truck, enabling the team to lift and lower the large storage container fitted to the truck bed. Also added were a refrigerator for storing scientific samples and a canvas roof over the passenger cabin.

New diving suits, constructed of lightweight material, were fashioned by U.S.D. Corp. (formerly U.S. Divers Company) in California. Designed to provide protection against such river hazards as piranhas and electric eels, they were made lightweight to keep the divers from becoming overheated—a critical factor in the tropics.

In December, Society researcher and writer Paula DiPerna was dispatched to the Amazon to begin the mammoth job of researching the expedition and establishing the logistical framework. (DiPerna would spend nearly a year paving the way for the Cousteau teams and then coordinating the movements of the many crews dispersed throughout Amazonia.) She set up an expedition headquarters in Manaus, established cooperative agreements with the Brazilian Amazon research center, Instituto Nacional de Pesquisa da Amazonia (INPA), and traveled widely throughout the Amazon basin doing advance work.

In January Captain Cousteau, Jean-Michel, Paula DiPerna, and Jacques Constans flew to Brasília, where they met with Brazil's President Joao Batista Figueiredo and various government ministers to obtain critical authorizations which would permit *Calypso* and her teams to enter and explore the Amazon. From the Brazilian capital they flew to Lima, Peru, and to Bogotá, Colombia, where they secured parallel authorizations for work in the western regions of Amazonia.

By the end of January 1982 the necessary permits had been arranged, the accumulation of jungle and river equipment was nearing completion, and the work on *Calypso* had reached the point of final touches. A date of February 27 was set for the ship's departure from Norfolk. Because the event would initiate a dramatic new chapter in the history of *Calypso*, a Society team set to work producing a documentary film about the unique preparation; it was called *Calypso Countdown*. Among the scenes is an inspection of *Calypso* by Captain Cousteau only a week before the scheduled departure: he meticulously explores the ship from the bridge to the engine room, checking out each new piece of equipment and every repair; when the day-long review is complete, Cousteau pronounces his ship ready for the Amazon.

February 27 arrived, a chilly, rain-soaked Saturday; a crowd gathered on the Norfolk dock alongside *Calypso*. Beneath a tent, Society staff members set out a five-hundred-pound bon voyage cake and, after a brief ceremony and a festive farewell to the crew, *Calypso* drifted backward into Norfolk harbor, turned seaward, and sailed out of sight into the Atlantic.

As *Calypso* steamed southward, an Andes land-exploration team was organized in Norfolk, to be directed by Jean-Michel Cousteau and Dominique Sumian. Cinematographers Jean-Paul Cornu and Louis Prezelin were assigned to this team, along with biologist Dick Murphy, sound engineer Guy Jouas, and film director Jacques Ertaud. The Land Team, after preparation in Norfolk, flew to Lima, then to Cuzco in the Andes, in the early spring. They spent a month studying the source of the Amazon, ultimately ascending Mt. Mismi, an eighteen-thousand-foot mountain. The snowmelt from Mismi engenders a series of small streams that become the Río Horillos then the Río Apurimac, then the Ene and Tambo and Ucayali, and, finally, swell into a broad river that is called Solimões by some until it merges with the giant Rio Negro near Manaus, at which point it is known to all the world as the Amazon. The Land Team traced the Amazon's headwaters by kayak and Zodiac, visiting Machu Picchu along the way, and arriving at a tiny port called Luisiana, where they inflated the huge raft and set out again.

Calypso, meanwhile, ran into a series of mechanical delays en route to Belém, the port city at the Amazon's mouth. An engine problem first caused the ship to pull into Savannah, Georgia, for repairs; then, nearing the Dominican Republic, the ship lost electrical power and its radios for a brief period. Next, as it approached Fort de France, Martinique, one of the two propeller shafts broke, forcing the ship into dry dock to replace the vital piece. Mysteriously, the shaft problems continued, with another breaking near Trinidad and a third, one of the new replacement shafts, breaking near Paramaribo, Surinam. Since no dry dock was available in that harbor, the ship limped back to Georgetown, Guyana, where the shaft was again replaced. A close inspection finally produced the explanation for the shaft failures. When *Calypso* had been emptied of all her contents in Norfolk, the old wooden hull sprang apart slightly. This warping effect was made permanent when a new bulkhead was built after installation of the new motors, causing a slight bend in the angle of the shafts between the transmission and the propeller; this distorting pressure caused the breaks.

Finally, on June 1, 1982, *Calypso* arrived to receive a welcoming celebration in Belém. As part of the ceremonies, a *carranca*—an Indian totem made of wood—was installed at the peak of the ship's bow, a unique kind of bowsprit that Brazilian Indians employ to ward off evil spirits in their canoes. Perhaps the magic was effective: *Calypso* experienced no more shaft problems as she left Belém and sailed, over the next three months, inland as far as Iquitos in northeastern Peru.

In Manaus, Brazil, *Calypso* was joined by a riverboat purchased for the expedition. With a draft of only two feet, the fifty-one-foot-long vessel could be used to penetrate far up tributaries, as well as providing additional lodging for as many as eleven people. It was dubbed the *Anaconda*.

Calypso and the Land Team made rendezvous near Iquitos, ending the first phase of the Amazon expedition. The work would continue through most of 1983, with teams dispersed into the uncharted wilds north of the Rio Negro and along the southern periphery of Amazonia, where a flood of poor settlers has been rapidly cutting and burning great swaths of the forest for development. Among the important studies to be made were those of water-related activities of Amazon Indian tribes, the amount of pollution in the river system, the nature of fish populations, and the characteristics and behavior of such interesting Amazon animals as dolphins, sloths, turtles, manatees, capybaras (the world's largest rodent), jacares (an alligator relative), anacondas, and more.

When finished, the complex *Calypso* expedition throughout Amazonia appears likely to produce the first system-wide water quality data, discoveries of new fish species, information on pollution "hot spots," and perhaps the most extensive documentary films about this vital ecosystem ever made.

And where to, after the Amazon? To the South Pacific, declares Captain Cousteau, and then to China. And so the legend continues—and the little vessel that engendered a universal appreciation for the beauty and the importance of life below the world's many water surfaces sails on.

Far from the heat of the jungle below, Jean-Michel Cousteau *(holding the flag)*, cameraman Jean Paul Cornu *(left)*, and their Peruvian guide celebrate reaching the source of the Amazon's waters some 18,000 feet up in the Peruvian Andes.

Index

All references are to page numbers; illustrations are indicated by *italic* type.

Photographs by the following appear throughout the book:
Raymond Amaddio, Dominique Arrieu, Claude Caillart, CEMA, Ron Church, COF, Anne-Marie
Cousteau, Cousteau Group, Cousteau Society, Scott Frier, André Laban, Jean Lattés, Jacques Lena, Bill
Macdonald, Richard Murphy, OFRS, Yves Omer, Lev Poliakov, Louis Prézelin, Jacques Renoir, Les
Requins Associés, Jacques Roux, Ludwig Sillner, Alexis Sivirine, Dominique Sumian, Paul Zeuna, and
other members of the Cousteau team.